DEVELOPMENT AS PROCESS

How can the complexity and unpredictability of planned development be understood?
How can project managers deal with the social relationships and institutional contexts in which they operate?

Linking practical experience and contemporary social theory, this book offers alternative ways of thinking about 'development as process' and new methods for field research and programme monitoring. Conventionally, the complexity of development work has been 'managed' through the use of simple project models in which planned inputs lead logically to predictable outputs. In focusing on the unintended outcome, the unmanageable element, the local variability of effects, and the importance of social relationships, the contributors in this book challenge simplistic managerial models and suggest new approaches and methods which acknowledge, explore and positively engage with the unexpected and with diversity in the development process.

Drawing on work in agriculture, irrigation, forestry, and fisheries in countries in Asia and the former Soviet Union, *Development as Process* examines changing information needs faced by development agencies as they shift from simple technology-led project approaches, towards an emphasis on policy change, institutional reform and inter-agency partnerships. In looking critically at the politics of information production and use in different cultural and institutional settings, *Development as Process* goes beyond method and technique and proposes a new look at the role of monitoring information in planned development.

David Mosse is Lecturer in Anthropology at the School of Oriental and African Studies, London. **John Farrington** is Research Fellow and Coordinator of the Natural Resources Group at the Overseas Development Institute, London. **Alan Rew** is Professor of Development Policy and Planning and Director of the Centre for Development Studies at the University of Wales, Swansea.

ROUTLEDGE RESEARCH/ODI DEVELOPMENT POLICY STUDIES

1. NATURAL RESOURCE MANAGEMENT AND
INSTITUTIONAL CHANGE
Diana Carney and John Farrington

2. DEVELOPMENT AS PROCESS
Concepts and methods for working with complexity
Edited by David Mosse, John Farrington and Alan Rew

Also available from Routledge:

IMF PROGRAMMES IN DEVELOPING COUNTRIES
Tony Killick
PB 0–415–13040–9

IMF LENDING TO DEVELOPING COUNTRIES
Graham Bird
PB 0–415–11700–3

MANAGING WATER AS AN ECONOMIC RESOURCE
James Winpenny
PB 0–415–10378–9

DEVELOPMENT AS PROCESS

Concepts and methods for working with complexity

Edited by David Mosse,
John Farrington and Alan Rew

London and New York

First published 1998
by Routledge
2 Park Square, Milton Park, Abingdon, Oxon, OX14 4RN

Simultaneously published in the USA and Canada
by Routledge
270 Madison Ave, New York NY 10016

Transferred to Digital Printing 2006

© 1998 David Mosse, John Farrington and Alan Rew, selection and editorial
matter; © 1998 contributors, individual chapters

Typeset in Sabon by Routledge

British Library Cataloguing in Publication Data
A catalogue record for this book is available from the British Library

Library of Congress Cataloging in Publication Data
Development as process: concepts and methods for working with
complexity/edited by David Mosse, John Farrington and Alan Rew. Includes
bibliographical references and index.
1. Economic development. 2. Economic development projects. I. Mosse,
David. II. Farrington, John. III. Rew, Alan, 1942– .
HD75.D4849 1998
338.9——dc21
97–32107
CIP

ISBN 0–415–18605–6

CONTENTS

CONTRIBUTORS

Ruth Alsop has a PhD from the University of East Anglia focusing on the social implications of Indian agricultural research. She was Programme Officer for Sustainable Agriculture with the Ford Foundation in Delhi until 1995, and has subsequently been made Visiting Research Fellow at the International Food Policy Research Institute. She is also affiliated to the Overseas Development Institute. She is currently researching the concept of social capital in the context of public sector responses to demands and options for technical change in agriculture in Rajasthan.

Angelika Brustinow is Lecturer in Development Studies, University of Wales, Swansea specialising in Bengali studies and in social development in the former Soviet Union. She has studied in Russia and Germany and has extensive project and policy operational experience in Bangladesh and Russia, Ukraine and Central Asia mainly for the Know How Fund. Her email address is a.brustinow@swansea.ac.uk

Anne Coles is a Senior Social Development Adviser in DFID's Social Development Division. She has been with DFID for five years. Her MSc is in human nutrition and her PhD is in human geography. She has many years' experience of development issues and has combined practical project implementation and monitoring with consultancy and postgraduate teaching in development studies.

Rick Davies is a Social Development Consultant, and Research Fellow associated with the Centre for Development Studies, University of Wales, Swansea. His research is centred on organisational learning, and his consultancy work on monitoring and evaluation, especially participative and interpretive approaches. He also manages Mande NEWS, a web site focusing on these issues, on behalf of four UK NGOs. He can be contacted at rick@shimbir.demon.co.uk and his web page can be found at http://www.swan.ac.uk/cds/rdl.htm

Phil Evans is a Social Development Adviser in the Department for International Development. He has been working in international development for fifteen years, after completing a PhD in social anthropology in 1983. He has recently been employed in DFID's Evaluation Department, having previously worked in the British Development Division in Eastern Africa, and the West and North Africa Department. In January 1998 he moved to DFID's Social Development Division as Senior Social Development Adviser.

John Farrington has a PhD in agricultural economics from the University of Reading, and is currently Visiting Professor there, as well as being Coordinator of the Natural Resources Group of the Overseas Development Institute. His main research interests are in the prospects of pluralistic approaches to technical change in agriculture and natural resources management.

Elon Gilbert has a PhD in agricultural economics from the University of Stanford. He has worked extensively in sub-Saharan Africa and South Asia on the provision of agricultural services. He has worked in Rajasthan with the Overseas Development Institute since 1994 on pluralistic approaches to the provision of agricultural research and extension.

Charlotte Heath is a Social Development Adviser in DFID in the West Asia Department covering Pakistan, West Bank and Gaza, Jordan and Northern Iraq. Areas of particular interest include decentralisation and community participation, people's participation in the analysis of poverty and policies to combat it, and gender equality.

Sudarshan Iyengar has a PhD in Economics from M.S. University, Baroda. He has worked in the areas of decentralised planning, irrigation economics, involuntary displacement and rural development. His current interests include action research on involuntary displacement in development projects, NGOs in development and issues relating to natural resources, poverty and environment. He has co-authored two books and contributed to books and journals.

Rajiv Khandelwal has a postgraduate qualification from the Institute of Rural Management, Anand, India. He has worked since 1987 on projects relating to agriculture development, ecodevelopment, education and environmental action with a number of NGOs in Rajasthan. Since 1994 he has been affiliated with the Overseas Development Institute in its study of pluralistic approaches to the provision of agricultural research and extension.

David J. Lewis is Lecturer in Non-Governmental Organisations at the Centre for Voluntary Organisation, London School of Economics. He was previously a Research Associate at the Overseas Development Institute and has undertaken research and consultancy for a number of governmental and non-governmental organisations in South Asia. His most recent book is *Anthropology, Development and the Post-Modern Challenge* (1996) written with Katy Gardner.

David Mosse (email: dm21@soas.ac.uk) has a DPhil from Oxford University and is currently Lecturer in Social Anthropology at the School of Oriental and African Studies, University of London. He has worked as Oxfam Representative for South India and as a long-term consultant on participatory natural resources development projects in India. He is currently undertaking anthropological, historical and process documentation research on indigenous irrigation and water resources management in southern India.

R. Parthasarathy (PhD) is an Associate Professor at the Gujarat Institute of Development Research. He has carried out research on labour use in agriculture, rural labour markets, the diffusion of agricultural technology, NGOs, natural resources and tribal economy. He has co-authored a book and has published a number of research articles in journals. He also works as a development consultant. Currently, he is engaged in documentation of the implementation process of the irrigation management 'turnover' programme in Gujarat, the use and management of natural resources, the role of Non-Government Organisations and people's participation in development programmes. He can be contacted at the Gujarat Institute of Development Research, Gota – 382 481, Ahmedabad, India.

Alan Rew is Professor of Development Policy and Planning and Director of the Centre for Development Studies in the University of Wales, Swansea. He has widespread international experience of process planning and monitoring especially in the Pacific, South Asia and East and West Africa. He is currently researching poverty, rural livelihoods and social capital formation in the context of the Chotanagpur Plateau or Jharkhand area of India. He also maintains strong teaching, research and operational interests in social policy and access to institutionally allocated social provision. His email address is a.w.rew@swansea.ac.uk

PREFACE

The idea of producing a book on process approaches was born in the aftermath of an informal workshop jointly organised by the Overseas Development Institute (ODI) and the Centre for Development Studies and held at ODI in April 1995. The workshop, entitled 'The Potential for Process Monitoring in Project Management and Organisational Change: Lessons from the Natural Resources Sector', was attended by practitioners, academics and policy researchers, mainly from the UK, but with some representation also from France, the Netherlands and the Philippines.

In many ways, the workshop was exploratory: some came with the notion that process approaches were important, but without direct experience of them; others came dissatisfied with conventional, deterministic project approaches, but at the same time uneasy over participatory rural appraisal (PRA) and other 'rapid' methods which many advocate as alternatives; yet others came to share the experience they had gained in introducing approaches more flexible, responsive, and sensitive to idiosyncrasy than conventional project design.

Continuing interaction with the workshop participants and others, together with the editors' long-term involvement in the study and application of process approaches in southern Asia, the former USSR and elsewhere, made it clear that a volume providing practical examples of process approaches and locating them within a conceptual frame would be welcomed by many.

This book aims to meet that need. It opens with a conceptual chapter by Mosse, followed (Chapter 2) by his overview of the case study material presented in subsequent chapters. Chapters 3 and 4 provide examples from India and Bangladesh of process approaches to information and monitoring in development projects. For the same two countries, Chapters 6–8 set out process experiences in the context of multi-agency collaboration. Chapter 5 sets out the expectations and practice of the UK Department for International Development in relation to monitoring 'process projects'. The final chapter, taking examples from Indian forestry and Russian land privatisation, examines how process approaches can influence policy reform.

ACKNOWLEDGEMENTS

The editors would like to thank the Natural Resources Policy and Advisory Department of the Department for International Development (then the Overseas Development Administration) for its financial contribution to the costs of the April 1995 workshop and to certain costs associated with the preparation of this volume. The opinions and interpretations presented here are, however, those of the authors alone.

INTRODUCTION

1

PROCESS-ORIENTED APPROACHES TO DEVELOPMENT PRACTICE AND SOCIAL RESEARCH

David Mosse

Introducing process monitoring and documentation

This book deals with some new approaches to information management in development programmes referred to as 'process documentation' or 'process monitoring'. While these approaches are intended to complement rather than replace existing routine monitoring and participatory appraisal methods, the concept of 'process' itself implies a different perspective on development initiatives. This chapter will explore this 'process' perspective as well as showing why it is important particularly in view of recent changes in the nature of development initiatives themselves. This means identifying some limitations presented by conventional information systems in development and the reasons for wishing to identify different methods. In the second part of the chapter, I will set process monitoring and documentation in the context of various other social science involvements in development practice over the past two or three decades. This will set the scene for the examples of process documentation and process monitoring discussed in later chapters of the book. In short the key questions are: what does it mean to view development as a 'process'? Why should we monitor or document this process? What approaches are there to understand this, and how do they differ from existing methods in social science?

Although this chapter discusses general features of 'process' monitoring and documentation, it must be stressed at the outset that the meaning and purpose attributed to this concept will vary depending upon the point of view involved. For example: a policy adviser may view process monitoring/research as a means to investigate the working and viability of a generally applicable development model (e.g. privatisation or resources management transfer); a project manager may view it as a means to monitor

3

progress in implementation, to generate solutions to a specific problem, or to justify decisions; an academic researcher may be more interested in developing a new interpretative view rather than solving a particular problem; a funding agency may view process monitoring and research as a means to improve inter-agency collaboration within a 'managed network', while members of the network themselves may be more concerned with self-regulation, communication and establishing the acceptability of new ways of relating. A local NGO may see process monitoring as a means to explain programme effort and impact to outsiders and so to mobilise funds or government support, while a group of villagers may view process monitoring as a means to record a significant struggle, perhaps one which led to a positive redefinition of group identity or to significant change in local social relationships.

It is important not to submerge these different interests, perspectives and types of work in the creation and reification of a 'new' development buzz word or package of methods (and for this reason I choose *not* to adopt a convenient summarising acronym – e.g. PMD or PMR). Different interests imply different approaches, methods and interpretative frameworks. It is important that the conception of process monitoring/documentation is allowed to retain these different (even conflicting) interests. It must, for example, retain a link both to the pragmatic concern with more effective development practice *and* the concern for broader reflective understanding even though these may at times be incompatible orientations. I will return to this last point.

What does it mean to view development as a process?

In a literal sense development 'process' means concern with the 'progress' or 'course' of a project. It describes the actions and events arising from planned inputs and the means by which outputs are produced. This conception is close to another meaning of the term 'process', namely as a series of operations in manufacture (see Rew and Brustinow, this volume). This marks an important shift away from the focus on project inputs and outputs and the assumed mechanical link between them. Indeed, as a descriptive metaphor for development initiatives, 'process' is increasingly used as an alternative to the machine metaphor. Like other commonly used metaphors (including 'development' itself) the concept of 'process' provides a device for thinking and talking about a complex social reality in new ways (cf., Alvesson 1993).

There are at least three distinct ways in which the process metaphor signals an alternative to conventional models of the development project. First, in relation to planning, viewing a project as a 'process' means having a design which is flexible and changes as a result of learning from implementation experience. This 'learning process' approach (Korten 1980) implies treating development projects as flexible systems with changeable procedures and approaches. As an ideal-type, 'learning process' is contrasted with the 'blueprint' approach in which a project is designed to be delivered in a specified form (known inputs, activities, outputs and costs) and to a fixed time-frame. Second, 'process' refers to the relationship and contextual elements in all projects. All projects, even those with 'blueprint'-type designs, have permeable boundaries and are influenced by their wider social and institutional environment.[1] Relationship elements have typically been under-recognised, treated (if at all) informally, viewed as a source of problems and misunderstandings rather than as an essential part of the development effort requiring explicit planning or managerial attention. Third, 'process' refers to the dynamic, unpredictable and idiosyncratic elements in development programmes; those things which are not easily amenable to planning and management control but which are nonetheless central to success or failure (Korten 1989; Uphoff 1992).

There has been a tendency too for conventional tools of programme planning and monitoring to ignore 'process elements' and to treat projects as closed, controllable and unchanging systems. In *planning*, for example, 'logical framework analysis' – a widely used planning device – involves deliberately isolating hypothesised causal links in order to determine predicable project outcomes. Elements of relationship and institutional context which are often fundamental to achieving success are treated as 'assumptions' largely falling beyond management control. Conventional *monitoring* (and evaluation) approaches (e.g. before/after studies) enable managers to know the extent to which programme inputs produce the predicted results. The approach is essentially deductive; or, as Korten (1989: 8) puts it, each project becomes a formal experiment, and monitoring and evaluation concern the testing of a given project hypothesis by measuring output and impact in terms of pre-defined indicators. However, this approach leaves little room to describe the unplanned impact or the unexpected change; nor does it help explain *why* or *how* particular outcomes were achieved (Korten 1989: 15). Such questioning of the causal assumptions that link inputs to outputs in project design is especially important where new approaches are being piloted for future expansion.

Why should we monitor or document this process?

Of course, logical frameworks and indicator-based monitoring systems are necessary tools of planning and management. Indeed, it is hard to conceive of purposeful, planned activity which is not based on hypothesised causal relations. However, common sense and experience tells us that the simple project model is dangerously far from reality; that the relationship between inputs and outputs is not linear; that responses to inputs are often non-proportional, that action generates unpredictable effects and that the same inputs under similar conditions do not always produce the same results

(Korten 1989: 15). Development thinking, Uphoff suggests, would do well to take less account of mechanical metaphors and more of 'chaos theory' which explains how small causes can have large effects, how 'a butterfly stirring the air today in Peking can transform storm systems next month in New York' (Gleik 1987: 8, cited in Uphoff 1992: 294). Development action is undeniably complex, often unpredictable and locally variable in its effects, and significantly influenced by realms over which management has limited or no control (e.g. culture, politics, institutions, policies, costs or prices). Many planners know from experience that social and political relationships involved in development settings influence outcomes as much as carefully designed inputs. There is, however, no means to use such knowledge in development planning which conspires, instead, to create imaginary and simplified planning worlds.

Concern with 'process' is not simply an intellectual fashion.[2] There are some pragmatic reasons why agencies are today more interested in process dimensions of their programmes than they have formerly been. The challenge to simplistic project models indicated above has been underlined by recent changes in the nature of development programmes themselves. These have made the limitations in existing planning and monitoring systems more visible. There are several interrelated shifts in approaches to planned development (evident from the early 1980s) which are worth mentioning.

First, there has been a shift away from narrow technology-led projects and a greater emphasis both on sectoral concerns (sector-wide reform or strengthening) and cross-sectoral issues (e.g. poverty, gender). Second, this has meant that the bounded project is no longer the exclusive focus of development assistance. Managed networks and inter-agency links and partnerships are increasingly important in meeting programme objectives which embrace wider goals of policy change or institutional reform. Development intervention is less and less captured or packaged in 'the project' as a closed controllable system (Korten 1989: 8). Indeed, there may be a more general shift from 'project-centred to organisation-centred concerns' (Marsden *et al.* 1994: 162). Third, there is often a move from externally planned, technically and managerially prescriptive ('blueprint') approaches in development planning, towards more flexible and iterative approaches, characterised by Korten (1980) as 'learning process approaches' in which neither means nor ends can be fully known in advance (Uphoff 1992: 12).[3] This comes out of the experience that development 'solutions' often evolve from experimentation and practice rather than from design. More precisely (following 'logical framework' terminology) while programme goals and purposes may be clearly stated, programme outputs and activities are devised (and revised) in the early stages of the project itself on the basis of project experience. Fourth, there has been a shift from centralised and 'top down' approaches towards more decentralised and participatory ones.

These changes in strategy themselves arise from past failures and new policy goals. They are, for example, a response to the high cost and poor performance of centrally planned technical projects, and the concern to reduce public sector costs and increase effectiveness and long-term sustainability of development interventions through the involvement of local people and non-governmental or private sector agencies. At the same time a changed macro-economic and political environment has generated new aid policy goals such as 'good governance' or 'political pluralism' which now foster formerly unlikely collaborations (e.g. government and private sectors, academic institutions and NGOs, between government departments and militant grass roots organisations (e.g. in the Philippines)).

These changes have not only drawn attention to neglected 'process' aspects of projects, but have also created a demand for information in new forms. More open-ended project designs and the piloting of approaches and institutional arrangements, for example, are premised on rapid information feedback and learning from practice. Process and participatory approaches recognise that different stakeholders have different interests and that their ownership and commitment is important to achieving successful outcomes. This gives a new emphasis to the understanding and monitoring of institutional interests and relationships, and to inter-agency communication and consensus building (i.e. making learning a *joint* process). This emphasis on relationships is all the more important given that projects are less and less about creating deliverable products (e.g. bridges, power plants or other productive technology and services) and more concerned with introducing behavioural changes which have to be sustained in the longer term. The explicit concern with institutional development and sustainability has itself resulted in an increase in institutional complexity in development. Donor managed projects and bilateral links between donors and technical line departments have given way to multi-agency partnerships and collaborations (e.g. between international donors, GOs, NGOs, the private sector and research centres) many of which are as yet exploratory, experimental or politically sensitive.

These new approaches and complex policy agendas imply a need for kinds of information generation and communication which differ from those that satisfied simpler project models in the past. Institutional change produces 'outputs', for example, in forms which are difficult for conventional monitoring systems to recognise. Moreover, unlike physical outputs (e.g. infrastructure, technology uptake) new institutional arrangements (groups, links, forums and the like) cannot be judged from a universal, professional vantage point. Monitoring systems need to take account of the different points of view of different 'stakeholders' (e.g. providers or users of a service) who have socially determined perspectives on, and criteria for, institutional performance.

If the monitoring of social and institutional performance is complex, the

7

monitoring of social and institutional *impacts* – enhanced capacities, empowerment, organisational change, attitude and value shifts, public policy impact, etc. – is even more so.[4] Such changes are not easily predicted or managed. Indeed, attempts to monitor institutional change within conventional monitoring systems often generate absurdly large numbers of indicators. Clearly, there is need for non-predictive, non-indicator based systems of institutional monitoring which can identify and feed back information on significant changes generated by programme activities (Davies, this volume). Such approaches are inevitably inductive, selective and inter-pretative.

Awareness of these new challenges has led to experimentation with the various process-oriented methodologies discussed in this book. Most are geared towards monitoring programmes *as they occur* in specific contexts and feeding back information which can help managers, researchers, policy makers or network members respond to events. As some of the case studies will show, the availability of information on 'process' can serve to analyse failure, adapt approaches and in other ways facilitate more rapid (manage-rial) responses to events, lead to 'course corrections' (Lewis, this volume), and stimulate modification of project objectives and strategies in the light of implementation experience.

In addition, there are several more specific and different purposes for which process information is sought. First, it may be viewed as a means to develop agency capacity to undertake a *new* and complex task (e.g. the promotion of collective action for resources management) (see Chapter 2). This may involve the refinement of operational procedures prior to the adoption or expansion of a pilot programme. Equally, more open systems of information exchange (through wider stakeholder participation in moni-toring) may contribute new ideas for programme modification. Second, and relatedly, process information may provide the means to *validate* a new approach, to lobby for policy change (Rew and Brustinow, this volume) and inform the design of future projects. Third, process information may be viewed as a means to explain impacts and to produce promotional material for an agency (e.g. impact stories for donors). Fourth, process monitoring is aimed at understanding *inter-agency collaboration*, to analyse how partner-ships work and to assess an impact of collaboration on organisational performance (a donor agenda). By focusing on inter-agency interactions and exchanges, 'process monitoring' can help determine the conditions necessary for effective collaboration (e.g. between government, NGOs, and the community), or test the assumption that such collaboration achieves economies, and enhances performance (Gilbert *et al.* 1995).

Fifth, process monitoring has been used to construct a critical 'institu-tional ethnography'. That is, to analyse prevailing discourses, consensus models (e.g. of partnership, or community) and to identify underlying insti-tutional objectives which may mis-direct work or mis-specify problems.

Process approaches to monitoring provide new analytical opportunities for broader understanding and critical feedback. They may, for example, expose unstated individual and organisational objectives, which are concealed behind 'official' consensus views and project models, but which have a decisive (positive or negative) influence on the course of a programme; or indicate areas of conflicting interest or mis-communication. A critical 'archaeology' of projects (Lewis, this volume) also helps to place development initiatives within their larger social and political context, to understand how choices and decisions have been made or pre-figured, which representations gain dominance, which voices become muted, and what external pressures drive development actions. Process research, then, provides instruments for policy research serving purposes beyond the immediate concerns of project management, such as the critical review of programme choices, organisation structures and development models.

Finally, 'process monitoring' is used as a means for *engagement* in institutional processes of negotiation and consensus building within programmes; it is the means to produce rather than record outcomes. 'Process monitoring' may, for example, be used to promote inter-agency understanding through the provision of a range of 'communication services' (Farrington, Gilbert and Khandelwal, this volume). The very complexity and diversity of programmes and agencies generates conflicting objectives and uncertainty about benefits (Rew and Brustinow, this volume). These could easily become the rocks on which development programmes are wrecked. Given multiple perspectives and agendas, the task of monitoring is no longer simply to manage impacts or outcomes. Rather it must play a major role in creating a framework for negotiating common meanings and resolving differences and validation of approaches (Rew and Brustinow, this volume). The role of process monitors is then more of advocacy, facilitation or nurturing than analysis. Process monitoring helps provide a framework for negotiating meanings, agreements and validating policy or development approaches and resolving differences.

What approaches are there to understand 'process'?

There is a spectrum of research and monitoring activity ranging from 'in-house' feedback and reflection to focused empirical research which can only be 'crudely subsumed' under the concept of process monitoring and process documentation. These have different objectives and methodological orientations and, as I have just indicated, different objectives. Most intensive are methods usually referred to as process documentation research (PDR) which involve village-level participant observation and record keeping by trained long-term resident researchers. Field-level activities, meetings, negotiation, decisions and implementation problems are meticulously recorded. Less

9

intensive adaptations of PDR rely more on structured interviewing, the reconstruction of events and the use of existing documentary materials. Similarly, process monitoring may be undertaken through the diaries of project field staff rather than by specialist researchers; review workshops and verbal reporting may replace detailed documentation or the use of video and tape recordings. Newsletters, forums, or subject papers are also part of the process monitoring tool box. It hardly needs to be said that none of these methods are new in themselves, although the use to which they are put, and the means of selecting information to record are innovative.

These various different methods and issues are described in the following chapter and in case study pieces in this book. It may nonetheless be helpful to outline some general characteristics which, even if not always applicable, may help differentiate process-oriented from other approaches to monitoring or social research.

First, in contrast to both planning/design activities and *ex post* evaluation, process-oriented work involves *continuous* information gathering over a period of programme work.[5] Information on 'process' provides neither a 'snap-shot' view of a development intervention, nor a measure of progress against a fixed set of indicators. Rather it is concerned with the dynamics of development processes, that means with different perceptions of relationships, transactions, decision making, or conflicts and their resolutions.

Second, what distinguishes process-oriented monitoring is its orientation to *the present*; that is 'the intimate relationship with what is happening right now' (Gilbert, personal communication). Within the 'project' cycle the emphasis is on *implementation* rather than on planning or evaluation (although process work feeds into both).

Third, process monitoring is *action-oriented*. From one perspective this means that the outputs of process monitoring are, in the first instance, directed towards participants who are in a position to react to them through immediate action. Process monitoring is likened to assisting with the strategic and tactical adjustments during a football match (Gilbert, personal communication). From another perspective, the action focus is a methodological orientation. The premise here is that intervention and change make visible structures or forces underlying social systems which are otherwise invisible (Uphoff 1992: 299). 'It is said that if you want to know reality, you must try to change it' (Volken *et al.* 1982, quoted in Uphoff 1992: 275). From this perspective the significance of 'learning by doing' is not only found in the immediate utility of the information it generates, but in its capacity to produce a better interpretative social science of practice.

A fourth characteristic of process monitoring is that it is *inductive and open-ended*. This counters the more common tendency of action-orientation to involve a narrowing rather than a broadening of the frame of reference. The focus of process monitoring and documentation is not narrowed down to expected outputs or impacts, but includes an account of events and relationships and diverse impacts including those which fall beyond the project as officially understood. In this way process information helps break away from the image of development projects or programmes as closed, static, predictable and controllable techno-rational systems. It draws attention to the areas falling beyond management control which nonetheless have important influence on project success and allows critical reflection on the project's own definition of problems and solutions. In its focus on context and interpretation, process monitoring methods borrow from ethnography. An inductive approach also allows for the decisive influence of the *particular* – the improbable individual person, idea or event – as well as the more predictable influences of context, roles, interests and constraints on development outcomes (Uphoff 1992: 331–3).

Fifth, process monitoring and research is usually situated outside of project structures and the routine flow of programme activities and information. It often involves special (non-programme) staff or settings. Moreover, the information generated does not pass through the usual filtering and packaging involved in hierarchical organisations, and counteracts 'the very abstract and filtered form and sometimes perverted form' in which information appears at the top of the hierarchy and on the basis of which decisions are made (Boulding 1985: 29, cited in Korten 1989: 14). Korten goes so far as to say that process research provides a 'clear window into the rich detail of uncensored field experience' (1989: 15). However, a number of process-monitoring approaches firmly reject the idea of access to 'uncensored field experience'. Not only is there no *singular* experience to record (implementation 'reality' looks different for different actors – project staff, leaders, women farmers, etc.), but also information boundaries are often deliberately maintained, for example, by field workers anxious about performance evaluation from supervisors, or by officials engaged in chains of 'rent seeking'.

Sixth, recognising that monitoring information is highly charged with interests, a characteristic of some process monitoring approaches is the explicit recognition given to the different perspectives and judgements of 'monitors' themselves, treating these as important data in their own right. Davies (this volume) for example emphasises the inter-subjectivity of the monitoring process. His approach attempts to bring individual evaluations, selections and filtering involved into the public domain, rather than burying them deeper in the desire for objective reporting.

Finally, it needs to be emphasised that process-oriented approaches are not a substitute for other forms of monitoring, impact assessment or planning. Indeed, existing planning tools such as logical framework analysis or 'stakeholder analysis' help frame the concerns of 'process' monitoring. Moreover, process monitoring provides contextual clues for the interpretation of routine quantitative data as well as itself providing data for evaluative studies at various stages in a project's development.

Having outlined a few common orientations in process monitoring and research, it is necessary to underline some of the contrasts. First, approaches vary in the extent to which process monitoring/research is tied to programme action rather than independent of it. Process monitoring may be undertaken by agency staff as a form of self-reporting, or it may be separately organised and involve specialised process researchers from outside the agency. These different approaches are not always compatible. As Alvesson discussing organisational research puts it, 'the goal of promoting [programme] effectiveness tends to rule out complicated research designs and "deep" thinking, while the promotion of broad critical reflection presupposes that the [research] project is not subordinated to managerial interests' (1993: 6).

Second, process data is directed to different users: local participants, field staff, the project office, the agency, the donor, the wider policy making community, etc. These need not be mutually exclusive. Process data may feed into managerial decision making, feed into wider policy formulation, and contribute to research agendas (e.g. Salmen 1987). Indeed, process monitoring work often strives to be multi-directional providing resources for managers, researchers *and* local participants. However, the potential conflicts involved mean that work is usually oriented primarily to one or two users of data (Veneracion 1989: 107).

Third, there is variation in the focus of work. Process information may focus on field-level implementation and intra-community relations, on the link between development agencies and local communities, on inter-agency or agency-state relations. Fourth, there is variation in the intensity of the work. For example, process research may either involve long-term, open-ended, participant observation by full-time field researchers, or irregular field visits using checklists, interviews, secondary materials (e.g. minutes of meetings) and review meetings to reconstruct field processes. Fifth, data may be more or less systematically and formally recorded in field notes and diaries. Occasionally reporting is largely verbal and only occasionally summarised and disseminated.

Different process monitoring approaches need to be used selectively, the type and timing of work being dictated by objectives, circumstances, and the type of development work involved. However sensitively undertaken, process-oriented monitoring and research involves tensions – between engagement and detachment, insider and outsider, action and reflection, practitioner and researcher, support and criticism, management and field, etc. While overall the forms of 'process' monitoring discussed in this volume are firmly oriented towards improved effectiveness in development, we should not rule out the broader reflective research agenda. But as the case studies will show the tensions involved cannot always be resolved.

Social science and development practice

The previous section has introduced the idea of process-oriented monitoring and research, and in Chapter 2 a number of examples will be given in order to clarify some alternative approaches and methods. The purpose of the present section is to view process-oriented research/monitoring methods in the context of a range of engagements of social science in development practice. It will be suggested that process monitoring and research represent new and potentially fruitful opportunities in the evolving relationship between social science and development practice. In particular, while recent trends in development have created the need or opportunity for new social science involvements in development, changes in social science (and anthropology in particular) mean that there may be a new convergence of interest between social research and development practice (see Veneracion 1989; Uphoff 1992).

However, it must first be recognised that there has for long been a fairly clear divide between the use of sociological knowledge *in* development practice and the sociology *of* development (i.e., the critical analysis of planned change), and a corresponding divide between applied and academic anthropologists or sociologists. I will briefly look at approaches and interests in each of these streams in order to provide the context for emerging methods of development process research.

Social research in development projects

There have been different moments in the use of social analysis *in* development projects. First, and perhaps still most commonly, social science research is associated with project monitoring and evaluation (M and E) and with orthodox M and E methodologies which emphasise 'the use of large scale sample surveys of project impacts on beneficiaries and the statistical manipulation of data gathered' (Hulme and Turner 1990: 173, citing Casley and Lury 1982). Social research became institutionalised within projects in this form in the 1960s and 1970s in the shape of specialised monitoring and evaluation units staffed by full-time professionals within bureaucracies and projects (Hulme and Turner 1990: 173). Although still the most common form of social science research *within* projects, the contribution of this evaluative research to project management and decision making has been limited. Orthodox survey methods are now subject to much criticism. Complex data sets and analysis produce information which is often inaccurate, unreliable, too complex to analyse rapidly and results which are available too late to influence management decisions (Chambers 1983). In consequence, M and E units often become self-perpetuating information producing systems marginalised and isolated by management, which makes its decisions on the basis of short field reports and supervisory visits (Hulme and Turner 1990).

13

A second way in which social analysis was drawn into development projects (particularly donor-funded ones) was through anthropologists hired as consultants. However, until the mid-1980s social science analysis occupied a peripheral position within development contexts which were strongly led by technical agendas. As Rew's analysis of documents concerned with project appraisal and evaluation points out, the conception of projects 'starts from an engineering or economistic discourse that makes issues of social agency and cultural identity only incidental to the project design and implementation' (Rew 1997, cited in Grillo 1997). Anthropologists were hired as problem solvers, and the 'social' issues they were called on to address were often conceived in terms of the explanation of otherwise unaccounted for failure, the analysis of residual 'cultural factors' or the management of risky action (e.g. forced resettlement). Terms of reference and report sections focused on isolated moments of failure, on social constraints, social risks and problems (displacements, beneficiary non-cooperation, the lack of up-take of innovations, etc.). Of course, anthropologists had offered similar support to colonial administrations in earlier decades of this century.

Initially, explanations of 'social constraints' had demanded knowledge of local contexts and the anthropologists/sociologists recruited to undertake detailed qualitative studies were able to offer broader and more contextual understanding of development processes. However, management and operational time-scales pushed such longer-term work to the margins. Anthropologists who wished to continue in operational work and pursue professional careers in advisory or consultancy work had to learn to work with shorter time-frames in widely differing social contexts. This work, backed by new 'social development' advisory and support units within donor organisations, resulted, by the mid-1980s, in some cross-cultural generalisation about 'sociological variables' in rural development (e.g. Cernea 1991). Anthropologists working in development consultancy found themselves involved in the production of decontextualised accounts of social processes in countries they knew little about. Ethnographic knowledge gave way to a new practice-based quasi-theory: checklists, rules of thumb and the desire for generalising theory and analytical tools to compare with those of economist colleagues. Indeed, generalising theory has increasingly come to be provided by game theory and institutional-economic analysis. The focus on policy (rather than disciplinary concerns) also demanded other shifts, for example: from social analysis which was descriptive to that which was (in varying degrees) prescriptive; from inductive to deductive approaches; from a focus on present or past forms of social order to the anticipation of and influence over future outcomes; and from the role of outside project evaluator to planner or manager.

The shift towards predictive modelling and planning has been accompanied by a third trend in social research within development projects. Social science inputs are increasingly set in the mould of a new discourse of 'participation'. In other words, social concerns in planning and design are increasingly defined in terms of people's participation (rather than risks and impacts). For the time being anyway 'participation' provides a location for anthropological concerns and expertise in the development policy map. In doing so, the concerns of sociological analysis have become fused with those of the quite different tradition(s) of popular mobilisation. The conjunction of participatory goals and the use of social science in planning has given anthropologists/sociologists a role in a large range of development initiatives aimed at promoting community management or 'building local institutions'. Whether in forestry, agriculture and watershed development, or urban programmes, there is a strong emphasis on the planning and initiation of community based organisations (CBOs) – water users' associations, village forest committees, micro-watershed groups, or savings and credit groups. Policy initiatives for resource management 'transfer' (i.e. from state to community) place considerable faith in the ability of such community institutions to improve the delivery of services, reduce costs, and to improve maintenance and the management of assets. Social scientists have, for their part, been given roles as 'architects' of community, in the design, promotion and support of such new community structures.

The three trends noted above – the frustration with conventional survey research, the re-orientation of social research towards planning, and the linking of social research to participatory goals – have generated two new developments. On the one hand, the managerial need to compress and rationalise learning has promoted rapid appraisal methods, mostly linked to planning. On the other hand, there has been interest in participatory and action-oriented methods of research and planning. These twin influences have driven the generation and popularity of participatory rural appraisal (PRA)[6] as a planning and research method.

PRA methods have been imaginatively and successfully employed in a wide range of development settings. Indeed, during the last five or six years they have gained a deservedly central position as methods for participatory planning, monitoring and evaluation. Given this prominence, and given the mistaken tendency to view PRA as *the* most appropriate method of information generation and analysis for all levels and purposes, it is important to examine some of the constraints to PRA, and to establish ways in which 'process-oriented' methods differ from and may have complementary advantages over PRA.

There are several aspects of social life which PRA cannot adequately reveal (Hinton 1995). For example, while PRA has often proved very effective at generating agro-ecological and economic information, it has not in practice provided particularly good instruments for the kind of analysis of *social relationships* which projects require: information on patterns of dominance and dependence, credit relations, factions and spheres of political influence and patronage, disputes, or even social relations within a project team. The reasons for this are analysed elsewhere (Mosse 1995c). Some of them relate to the particular techniques and forms of diagrammatic representation in use, but often more important is the social context of many PRAs. PRAs are often *public* activities, in the sense that (a) they are community or village-wide events, (b) they take place in the presence of persons of authority, (c) they involve representation of local conditions and needs to resource-bearing outsiders, and (d) they are directed towards planned community action (Mosse 1994; Hinton 1995; Pottier and Orone 1995).

Participation in these events and the consensus outputs they produce are determined by local social relations, which may give privilege and authority to certain opinions, priorities and perspectives while muting others. Public discussion may also follow given ground rules of safe discourse or express official models and so cover over significant social (e.g. gender-based) tensions or differences of interest (Mosse 1994; Pottier and Orone 1995). PRA events are often more likely to obscure than reveal the local social relations which shape them – 'the micro-politics of rural consensus formation' or the hidden agendas which underlie apparently spontaneous discussions (Richards 1995: 15, 41). Indeed, a good understanding of local configurations of power – local leadership styles, factions and alliances and gender relations – is a *pre-requisite* for the organisation of effective community-based PRA, and for the interpretation of its outputs. Also a deeper day-to-day familiarity with communities is often necessary to encourage disclosure on more sensitive issues (Hinton 1995).[7] The repeated use of PRA methods may not give project managers adequate understanding of the local worlds they are trying to change, or the differences between their project's own categories of institution and action and those of 'local' people. Consequently project assumptions often remain unchallenged. In short, PRA has evoked the anthropological interests in 'local knowledge' without, however, including the concern with the social construction of this knowledge (see Mosse 1996).

Significantly, it is through the participant observation of PRA events and their contexts (rather than the direct application of PRA techniques) that some elements of local power relations underlying knowledge production can be observed. Thus, the best material for social analysis may not be found in the consensual output of chart, map or diagram, but in the absences, the gaps, the corrections, disagreements or conflicts, even the complete failure of the exercise (as described in Mosse 1994). This suggests the importance of participant observation (by development workers) and the review of project activity as a source of knowledge on social relations. It also suggests the need for detailed ethnographic case-study analysis 'to calibrate and validate PRA/RRA in specific cultural and political contexts' to assess the extent to which it is able to 'evade cooption by local politics' (Richards 1995: 15).

The observation and analysis of events can be an important source of social learning, which, moreover, has distinctive research advantages. As Appadurai has pointed out, conventional interview-based research techniques (and, one can add, PRA methods too) usually attempt to capture the *outcomes* of events – the identifiable net outcomes of social processes (organisation and leadership structures, new linkages, input supply lines, community decisions, etc.) (Appadurai 1989: 271–2). However, many important social data are manifest not in the outcome but in the quality of transaction, in the *relationships* implied, in the aspirations and expectations as well as in the *post facto* outcomes (Appadurai 1989). The implementation of minor project activities, and the observation and recording of a sample of actual micro-events or transactions, can, in new ways, inform on the nature of social relations and power, bringing these into higher relief. In essence this is one objective of process monitoring and research.

In the KRIBP farming project in tribal western India, for example (see Chapter 2, and Mosse 1995c), observations on patterns of participation and the success and failure of activities helped to reveal the social identity of prominent actors, the attributes and dynamics of power and influence in villages. Patterns of participation were not only strongly determined by, but also *helped to reveal* local networks of influence. The implementation of small-scale activities highlighted the significance for planning of factors such as clan difference, religious difference, patronage, factional conflict, and leadership struggles (e.g. between statutory leaders and traditional tribal headmen). In this way critical reflection on project action generated knowledge about social relationships which was not easily accessible through conventional interview methods, or those of rapid appraisal (RRA/PRA). The external interventions penetrated the community projections of unity, revealing some of the underlying local political dynamics. This enabled more sensitive (and ultimately more successful) strategies for intervention and institutional development to be formulated and put into practice.

These and other experiences (e.g. *PLA Notes* No. 24) point to the complementary need for more detailed, informal and longer-term exposure to 'local' society to learn about relevant complexities and social differentiation which may be obscured in public PRAs. Attention is therefore being directed towards additional approaches to rapid and contextual learning (see Chapter 2) which also avoid the problems of large and cumbersome questionnaire surveys, but which have been somewhat overshadowed by the rising popularity of PRA as a method of information generation within projects. One of these is the approach of 'participant-observer evaluation' developed in 1982–4 by the anthropologist Lawrence Salmen who worked in urban Latin America (Salmen 1987). Salmen used ethnographic methods (residence and participant observation in project areas for five months plus small, focused surveys) to identify contrasting perceptions of achievement between beneficiaries and project professionals in World Bank urban upgrading projects. The interpretations that project participants placed on events were then fed back to project managers. In particular, Salmen's work identified the role of community institutions, political party affiliation and leadership structures in controlling communication between the project and beneficiaries. This had been wholly overlooked by the technical focus of the project (as had the fact that benefits in one project were largely going to absentee landlords) (Salmen 1987, cited in Hulme and Turner 1990: 172). Salmen's findings resulted in the re-design of several projects. His methods were also transferred to other countries where 'evaluations' were successfully carried out by trained participant observers.

In conclusion, the strong emphasis on the use of participatory appraisal and planning techniques on the one hand, and the dominant generalised, predictive economic modelling of community relations, on the other, has pushed longer-term descriptive ethnographic approaches into the background (cf., Richards 1995). This has, if anything, further widened the rift between academic and applied anthropology. However, there are signs that difficulties in the practice of participatory approaches (as well as worries about generalised models of community and local institutions) may contribute to a revival of interest in some traditional anthropological concerns among development practitioners. Development workers are increasingly worried that inadequate attention is being paid to the social processes underlying participatory development and that in consequence rhetoric is running away from reality (e.g. Mosse 1995b). It is obvious to most development workers that programme activity takes place in particular social and institutional contexts; and yet the new 'participatory orthodoxy' provides few tools with which to understand the relations of power, such as dominance and gender, which set the limits and social conditions of participation itself whether in research, decision making or development action (Mosse 1994, 1995b; Long and Villareal 1994; Nelson and Wright 1995).

In short, development practitioners are increasingly finding it necessary to address concerns of context, power and social structure in order effectively to work in development projects. Rapid participatory appraisal methods do not provide all the tools necessary for this. Several practitioners have therefore come to view an analytical sociology *of* development as a necessary adjunct to working in and for development projects. At the same time there have been significant changes in the anthropological and sociological study *of* development, which suggest a new alliance of interests between academic research and development practice.

The sociology of development

In the era of large quantitative evaluation studies (the 1960s and 1970s) the sociology *of* development was mostly concerned with the understanding of large scale systems of dependency and domination, driven by grand theories such as modernisation and dependency. At the same time, empirical micro-studies of development either focused on the local consequences of these wider economic and political shifts (e.g. analyses of social change in terms of concepts of 'tradition' and 'modernity'), or contributed to the ethnographic construction of timeless traditional 'little communities' isolated from wider systems (Wright and Nelson 1995). Much of this, and subsequent, sociological, anthropological and historical research has remained conceptually (and often physically) inaccessible to development policy and practice, partly because it has been primarily informed by its own shifting disciplinary concerns.

There were nonetheless from the 1970s a number of studies which focused on the impact of development programmes – especially larger scale state interventions such as the community development programme in India, or 'Ujamaa' in Tanzania (cf., Long 1977; Dube 1958), and a few on the formation of state bureaucracies (Cohen 1980; Fallers 1974, cited in Wright 1994: 15). However, these studies were surprisingly few and far between,[8] and they did not, typically, focus on the processes of programme planning and implementation as social phenomena.

During the last two decades, a number of theoretical shifts within anthropology have made development institutions, policy and practice the focus of new attention. For one thing, the weakening hold of functionalist models has coincided with new interests. First, there is a greater awareness of the importance of wider political and economic forces, and of state interventions of all kinds in shaping the structure of 'little communities' (cf., Robertson 1984). Just as colonial ideology and administrative practice is now seen as having constructed much of what was (in the 1930s–60s) taken as 'traditional' society in Asia and Africa, so contemporary rural society cannot be analysed independently of the state and international development discourse.

Second, anthropological studies have begun to turn critical attention towards the sociology of development ideas and institutions in themselves. Schaffer (Clay and Schaffer 1984), for example, directed attention to the process of public policy practice as an important area of social research, and Wood (1985) pointed to the powerful role of labels in policy discourse. Several more recent studies influenced by Foucault's work have taken a new and critical look at planning discourse, 'development narratives' and the historically specific interactions between knowledge and power which accord validity to particular images of social reality and particular types of scientific knowledge to the exclusion of others (Arce et al. 1994: 152; Hobart 1993; Escobar 1992, 1995; Ferguson 1994; Long and Villareal 1994; Sachs 1992). Ferguson's study of aid donor (World Bank) discourse, for example, shows how, as a backdrop for a strong technical agricultural agenda, rural Lesotho (in the late 1970s) was represented as a traditional undifferentiated 'aboriginal economy' and peasant society in ways which ignored its status as a 'labour reserve' within the wider southern African economic region. Similarly, focusing on environmentalist 'narratives', Fairhead and Leach (1996) have examined the power of a 'scientific' discourse of degradation which systematically misread environmental change in West Africa, to construct landscapes and shape programmes on a large scale.

One of the central elements of the anthropological commentary has been to show how development regimes and the scientific and managerial paradigms they purvey underestimate and marginalise 'indigenous knowledge' (Croll and Parkin 1992; Hobart 1993). Such critiques may, however, have lost some of their force with the growing hold of participatory approaches to development which ostensibly give primary emphasis to 'people's knowledge' and bottom-up planning. Nonetheless, a few anthropological accounts have begun to show how 'people's knowledge' and traditions, or community management, are themselves socially constructed (or invented) within the new participatory orthodoxy (e.g. Mosse 1996, 1997a, 1997b). These accounts show how influential policy concepts of participation can also be seen as being strongly linked to institutional interests or as having deep historical roots in the concerns of colonial administration. In sum, in various ways anthropological studies look at the way in which development problems, solutions and strategies are constructed by institutions in culturally and historically specific ways (cf., Douglas 1987); and they point to the misrepresentation of indigenous experience involved in the persisting tendency of development discourses to construct simplified, universal and manageable social worlds.

Third, anthropologists have also begun to provide studies of the content of administrative practices and detailed historical and sociological descriptions of development settings. This has revived interest in the rather neglected 'anthropology of organisations'. A recent collection (Wright 1994) brings together some anthropological work on organisations and the culture of management regimes (see also Alvesson 1993).[9] The detailed anthropological study of organisations, in fact, has a long history. In the 1930s and 1940s American factory-floor studies examined problems, such as worker resistance to a company incentive scheme or the consequences of company growth or collective bargaining and strikes, largely from a management point of view (Wright 1994: 8). Then the Manchester shop-floor studies of the 1950s and 1960s used more *participant* observer methods to focus on workers' informal organisations, strategies, rationality and conflict. One thing these studies did was to distinguish formal from informal systems within organisations – the one inscribed into organisational charts, job descriptions, etc., the other describing the way in which individuals and groups relate to each other which is informed by links beyond the organisation. They also examined the culturally specific relationship between the two systems. For example, the penetration of the informal into the formal system might be viewed as a source of corruption in one setting but as a source of innovation in another (Wright 1994).

Unlike early works, more recent studies of organisations examine a wider range of perspectives (management, worker, etc.) and do not search for shop-floor 'small societies' in isolation from the wider economic and political processes and social structures. Indeed, more context-specific analyses seek to embed the workplace in its various social contexts, or to view formal work settings as only separated from informal systems and wider society by semi-permeable membranes (Wright 1994).[10]

Like other forms of organisation, development institutions and projects also work through informal systems and are embedded in wider social contexts. Organisation studies endorse the remark I made above that development projects are not bounded entities or static, equilibrium systems formed around uniform consensual goals and ideas, but rather processes of organising and making meaning (Wright 1994: 19). Anthropological understandings, moreover, see these processes as essentially *political*. Development projects are political systems in which different perspectives contend for influence. They articulate relations of power which make certain ideas, values, problems and strategies of action (i.e. certain forms of discourse) authoritative. The contest 'to assert definitive interpretations which produce material outcomes' (Wright 1994: 23) is seen as central to planning and implementation within development projects at all levels (donors, implementing agencies and local communities). The anthropological task is to identify the social structures and interests upon which organisational processes and 'culture' are predicated.

But development situations provide particularly complex research settings because of the wide-ranging links between 'local' and 'global' institutions and social worlds. The study of micro situations in the context of global systems has long been a problem in ethnography. Several micro-level actor-focused studies have now been undertaken within the framework of an 'interface analysis' proposed by Long and Long (1992). These studies examine the intersection of local communities and development interventions, highlighting some of the miscommunications involved. Like Salmen's 'participant-observer evaluations' they look at the meeting of the world of project beneficiaries and development project professionals, examining the different 'life worlds' of actors and the tactical and strategic manipulation of information and knowledge at their interface. They link empirical (ethnographic) work with the analysis of the wider configurations of policy and administrative practice (Long and Long 1992; Arce *et al.* 1994). Local people and development agencies are seen as political actors pursuing often different agendas, yet negotiating development outcomes in particular social contexts.[11]

The study of development interfaces and of rapid change represents a major departure from (and challenge to) the anthropological study of enduring coherent cultures. The experience of 'cultural incoherence' and innovation discredits earlier descriptions of formal order (Spencer 1989: 147). It is no longer possible to view ethnography as a cross section of the flow of events and expect to find enduring social or conceptual structures. Rather culture itself is to be viewed as 'a series of processes that construct, reconstruct and dismantle cultural materials' (Wolf 1982: 387).

Moreover, as those concerned with the politics of representation have pointed out, the anthropologist is firmly part of this process of creating culture. No longer justified as a value-free and objective observer, the source of politically neutral and authoritative scientific knowledge, the anthropologist now has to conceive of his or her relationship to the situation under study in different ways. At one level there is an awareness that anthropology itself – having as its conventional focus the study of subjected people – cannot remain blind to the shaping of its own theory by its link and service to the dominant western civilising mission (Wright and Nelson 1995: 44). As Pels and Nencel put it 'classical anthropology hid its political projects . . . with the cloak of the neutral and value-free study of cultural difference' (cited in Wright and Nelson 1995: 45). This raises questions about the conditions of production of anthropological knowledge and about its accessibility, relevance, and usefulness to the people being studied. As Spencer points out, the issue here is not so much the literary concern with anthropology as 'text' or the 'writing of culture' (cf. Clifford and Marcus 1986), but rather 'who does anthropology, in what context, and to whom are the results made available' (1989: 161).

If current concerns have moved anthropology away from 'the presentation of seamless, polished accounts of other cultures' in which the anthropologist-narrator is invisible and omnipresent (Spencer 1989: 154, 157) towards more explicit concern with different points of view, cultural incoherence, and the specificity and idiosyncrasies of the personal field work experience, then this comes close to the concerns of process-oriented research in accounting for development outcomes. While anthropologists are now concerned with how accounts of 'culture' (i.e. anthropological texts) are produced, process monitoring tries to understand the conditions of the production of outcomes of 'development', which is surely as incoherent and contested a notion as is 'culture'. There may be other parallel shifts which bring the practice of anthropology and that of process research and monitoring close. As some practitioners of participatory development have become more concerned with critical reflection on process, some anthropologists perceive a need for greater accountability to, and involvement of, local people in their own self-representation. In both cases the boundary between academic enquiry and purposive involvement has been blurred.

The boundary does nonetheless remain. Many anthropologists remain critically disengaged from development practice, viewing legitimate action more in terms of public education or political advocacy for the self-determination of the local populations with whom they are familiar (e.g. Escobar 1991; Ferguson 1994). Moreover, while anthropologists may be sympathetic to new participatory approaches to development, as Wright and Nelson (1995) point out, there are significant differences in purpose and theoretical grounding between anthropological 'participant observation' and the 'participatory research' involved in development projects. Anthropological research aims to explore social relationships and processes of dominance as the context within which knowledge production and development action take place. PRA-based participatory planning, by contrast, has the aim of building consensus and planning for collective action.

Process-oriented work attempts to manage a tension between the tendency towards participatory action and that of critical analysis. But there is a dilemma here. The more independent, critical and unrestricted by project management concerns the study of project processes is (the more anthropological), the greater the loss of practicality and legitimacy (cf., Alvesson 1993: 30). Conversely, the more the analysis is instrumentally focused and tied to project management concerns, the less interpretative power it will have (Alvesson 1993: 33). The researcher cannot deal with the question, 'what is this thing (project, programme, network . . .) and how does it work?' at the same time as providing an answer to the question, 'how can *we* make this work better?'. If these contradictions are not recognised there is a risk that the idea of process monitoring and research will end up promising more than it can deliver either analytically or practically.

Beyond knowledge as a public good

Insofar as process-oriented documentation and monitoring aims to challenge simplified project models and provide more relevant information for decision making, it faces an even more fundamental difficulty, namely that information generation and use is itself inseparable from specific interests. Information cannot any longer be viewed as a public good willingly supplied and used to improve decision making, or increase accountability.

First, it is not at all self evident that an increased availability of information will improve the quality of decision making and action. Indeed, the preoccupation with information may have less instrumental value than we assume. Perhaps, as Feldman and March argue, it derives instead from a distinctively Western cultural context in which 'information symbolises reason, reliability, security, even intelligence and is thus a matter of legitimation' (1981, cited in Alvesson 1993: 50). 'Using information, asking for information, and justifying decisions in terms of information have all come to be significant ways in which we symbolise that the process is legitimate, that we are good decision makers, and that our organisations are well managed' (Feldman and March 1981: 178). Such a belief often serves to obscure the complex politics of information production and use within development projects and programmes.

Second, within organisations information flows are in various ways controlled, filtered and regulated. Institutional interests conspire to decide which versions of reality are legitimate. Certain models of reality are autho-rised not because limited information prevents better alternatives, but because they give legitimacy to given patterns of action or existing structures of interests. Information is channelled 'up' the system, exchanged or concealed to reinforce models and protect interests. Of course the way in which this happens depends upon an organisation's particular incentive structure (cf., Wade 1993, cited in Baumann 1996). But very few, if any, organisations are conducive to the free flow of information. Openness and accountability are rarely in practice consistently promoted.

Where it falls within the domain of management control, process moni-toring is likely to serve to validate and enforce existing perceptions and assumptions and consensus models. This may even be its principal purpose. However, process information is often specifically understood as that which is generated beyond routine monitoring channels. As such it is almost always subject to suspicion. To the extent that process documentation or moni-toring seeks to modify existing information flows or create new ones, its legitimacy is likely to be called into question, especially if it brings to light conflicts, mismatched aims and objectives, or programme failures. Some instances of this will be discussed in Chapter 2.

One particular institutional need which project models have to meet is the need for simplicity. Since, as Luhmann suggests, administrative systems can only function if they reduce complexity (1967, cited in Baumann 1996), the increased complexity generated by process information is likely to be viewed as unnecessary, undesirable or as reducing manageability. Process informa-tion is often presented as offering little to advance management control over development. On the contrary it may make this goal seem even more ungraspable. Monitoring systems which admit to producing information which is only personal, localised and subjective are hardly useful. Development workers and managers are increasingly bombarded with disorienting statements that projects are ongoing processes, that knowledge is negotiated, reality is 'emergent', that achievements are not objectively verifiable, or that 'success' means nothing, and context everything, etc. (Marsden et al. 1994: 29, 156–8).[12] In short that monitoring systems produce little that can be of use anywhere. The response to this uncertainty is often to attempt to grasp back control over an alarmingly elusive process (but one which nonetheless involves concrete expenditure) with the instru-ments of management – logframe indicators, milestones, etc. – and to reassert the simplicity of the model.

Third, it is not only project authorities or 'management' who seek to control information. For actors at all levels, information can be understood as a private good, 'part of the agent's private endowment and an important source and instrument of power in economic transactions; for their own benefits, agents seek to influence the other's decision by hiding, partially revealing, distorting or manipulating the pieces of information relevant to them' (Baland and Platteau 1993: 16, cited in Baumann 1996). Given the private nature of much process information, certain types of process monitoring and documentation are likely to infringe on individual interests. There would, for example, be strong individual (as well as collective) resistance to any public exploration of the way in which politics and administration are linked in the everyday practice of development (as in Wade's (1985) study of the market for public office). Among middle-level field staff there are likely to be strong incentives to conceal information on the nature of relationships involving covert chains of 'rent seeking behaviour'. Equally, field workers deliberately withhold or distort information from senior executives or in other ways restrict the sharing of information which reveals poor performance. This is not only done to place their own work in good light but also to secure support from the communities in which they work. Indeed, field workers often act as communication 'brokers bridging the gap between village and project management, constructing information so that villager desires can be translated into activities acceptable in terms of project objectives' (Mosse 1995b: 12). The outcome of this strategic use of information is that in every development project information boundaries are deliberately and carefully maintained. Information gets lodged in different parts of an organisation, its flow is deliberately directed, guarded or restricted by individuals holding conflicting priorities (cf., Edwards 1994). By any reckoning this is a serious challenge to the notion of information feedback for improved decision making.

Fourth, although notionally 'independent', process monitors or documentors themselves inevitably operate within frameworks which align them to certain perspectives/interests rather than others. Because all information flows are nested in relations of power, process monitoring and documentation will be perceived by different actors in different ways. For many, because it involves deviation from the existing structures and flows of information in projects, process-oriented work will be viewed as threatening. However for some, process monitoring may offer resources to better exert influence. It may, in other words, contribute to various forms of advocacy and lobbying.

Overall, the point to stress is that the close involvement of process-oriented monitoring in project systems and its specific intention to generate information usable within these same systems, makes the link between power and knowledge – a preoccupation of post-modernist social theory – unusually clear. It highlights the fact that knowledge (in development) is made valid not by its relation to its object (its objectivity) and not by the consensus underlying its assertions, 'but by its relation to our pragmatic interests, our communal perspectives, our needs and our rhetoric' (Baumann, citing Cahoon 1996).

This is not to say that process documentation and monitoring are under-pinned by any one theory of knowledge. Approaches range from those which attempt some form of 'independent' and 'objective' recording of events or analysis of relationships, to those which have abandoned any effort at analytical description of development settings. In the latter instances, process monitoring is firmly embedded in agency action or inter-action. Its role is more to facilitate communication than to generate new analytical insights. Information in this case has validity not to the extent that it throws light on a development 'reality', but to the extent that it is instru-mentally useful to the actors immediately concerned. Process information is, consciously, a resource strategically generated, used (or withheld) by actors to serve specific interests. Process information is indeed unlikely in itself to have any beneficial impact on development actions. Its effect depends upon the links set up (e.g. through lobbying or collaboration) which allow its use to strengthen the hands of different players in negotiating alternative outcomes. In short, process information is a resource in the ongoing politics of development encounters; one which can also make actors more self-conscious of the underlying relationship between information and power.

Notes

1 This aspect of the 'process' approach is emphasized in an 'Issues paper on process projects in India' prepared by Ita O'Donnovan for an ODA Workshop (New Delhi, 5–7 October 1994).
2 Fashion has, however, played a part in the dissemination of ideas of 'process' in development. Uphoff (1992) and Korten (1989) among others are keen to describe the change from 'blueprint' to 'process' type models as a Khunian paradigm shift related to a move in other disciplines 'from relatively closed mech-anistic, and reductionist models to more open, contextual and integrative ones' (Uphoff 1992: 19). This is seen as having begun in the early twentieth century with the challenge which a new physics based on quantum mechanics presented to Newtonian physics. The notion that ideas rather than material interests influ-ence behaviour, that the social world is constituted by systems of meaning in fact goes back to Durkheim and can hardly be construed as a new post-positivist social science.
3 This shift should not be overstated. 'Process' approaches to projects often take place within larger government organisations (national and donor bureaucracies) whose procedures are still characterised by a rigid system of project approval, evaluation, financial control and fixed short-term (4–5 year) funding cycles.
4 Recently discussion has focused on such problems in the evaluation of social development projects (Marsden and Oakley 1990; Marsden et al. 1994; Padaki 1995).
5 This is generally the case, even though the balance between direct observation of events and retrospective reconstruction varies.
6 PRA refers to a set of methods used to generate information with and by local people. One distinctive characteristic is the use of diagrammatic and matrix

scoring methods which allow non-literate people to produce and analyse information on their livelihoods and experiences and to communicate this to outsiders. For an introductory account of these methods, see Chambers 1991.

7 For further critical reflections on the practice of PRA from anthropological perspectives see *PLA Notes* (1995).

8 In India, major national programmes of the 1970s and 1980s on rural credit (primary credit cooperatives), or poverty alleviation (e.g. IRDP) attracted little attention from anthropologists.

9 Just as anthropologists have focused attention on the study of organisations, so researchers working within management and organisation studies have begun to use qualitative methods or to undertake their own 'organisational ethnography' (Van Maanen 1979; Mouly and Sankaran 1995).

10 A focus on context has also followed the shift away from general explanations of behaviour in organisations based, for example, on individual psychology or class conflict.

11 Up to this point, I have mentioned a few relevant strands in the anthropology of development, drawing mostly on British, Dutch and American work. For a wider sampling of current work illustrating different European traditions and those from developing countries (especially India) within the anthropology of development, see Grillo 1997.

12 Such conclusions arise from an analysis which is incomplete. What is relevant is not that knowledge is negotiated, or that meanings and value are contested, but which individuals, groups or institutions are able to make their particular definitions of situations authoritative, and in doing this to redirect material benefits in their direction, and how they are able to do so. An anthropology of development processes is thus political. As a research tool some kinds of process work have potential as a method to expose the relations of power which underlie development processes (as was done in an earlier study, BRAC 1984). Process research has a role in identifying interests, exposing fictitious consensus, and, importantly, in developing strategies to intervene on behalf of the excluded. Insofar as these involve directed action on behalf of the poor, process analysis is able to serve interests and ensure a better focus to development action.

References

Alvesson, Mats (1993) *Cultural Perspectives on Organisations*, Cambridge: Cambridge University Press.

Appadurai, Arjun (1989) 'Small-scale techniques and large-scale objectives', in *Conversations Between Economists and Anthropologists: Methodological Issues in Measuring Economic Change in Rural India*, Delhi: Oxford University Press.

Arce, A., Villareal, M. and de Vries, P. (1994) 'The social construction of rural development: discourses, practices and power', in D. Booth (ed.), *Rethinking Social Development: Theory, Research and Practice*, Harlow: Longman Scientific and Technical.

Baland, J.M. and Platteau, J.-P. (1993) 'Are economists concerned with power?', *IDS Bulletin* 24, 3.

Baumann, Pari (1996) A review of the literature on information generation and exchange: implications for process documentation and process monitoring. Unpublished manuscript, Overseas Development Institute.

Boulding, K.E. (1985) 'Learning by simplifying complexity: how to turn data into knowledge', in K. Boulding (ed.), *The Science and Praxis of Complexity*, Tokyo: The United Nations University.

BRAC, 1984 (1980) *The Net: Power Structure in Ten Villages in Bangladesh*, Dacca: Bangladesh Rural Advancement Committee.

Cahoon, L. (ed.)(1996) *From Modernism to Post-Modernism: An Anthology*, Oxford: Blackwell Publishers.

Casley, D.J. and Lury, D.A. (1982) *Monitoring and Evaluation of Agriculture and Rural Development*, Baltimore: Johns Hopkins University Press.

Cernea, M.M. (ed.) (1991) *Putting People First: Sociological Variables in Rural Development* (2nd edn, revised and expanded), World Bank, Washington: Oxford University Press.

Chambers, R. (1983) *Rural Development: Putting the Last First*, London: Longman Scientific and Technical.

Chambers, M.R. (1991) 'Shortcut and participatory methods for gaining social information for projects', in M. Cernea (ed.), *Putting People First: Sociological Variables in Rural Development* (2nd edn, revised and expanded) World Bank, Washington: Oxford University Press.

Clay, E.J. and Schaffer, B.B. (1984) *Room for Manoeuvre: An Exploration of Public Policy Planning in Agriculture and Rural Development*, Rutherford: Farleigh Dickinson University Press.

Clifford, J. and Marcus, G. (eds) (1986) *Writing Culture: The Poetics and Politics of Ethnography*. London: University of California Press.

Cohen, R. (1980) 'The blessed job in Nigeria', in G.M. Britan and R. Cohen (eds), *Hierarchy and Society: Anthropological Perspectives on Bureaucracy*, Philadelphia: Institute for the Study of Human Issues.

Croll, E. and Parkin, D. (eds) (1992) *Bush Base: Forest Farm: Culture, Environment and Development*, London and New York: Routledge.

Douglas, M. (1986) *How Institutions Think*, London: Routledge.

Dube, S.C. (1958) *India's Changing Villages: Human Factors in Community Development*, London: Routledge and Kegan Paul.

Edwards, M. (1994) 'NGOs in the Age of Information', *IDS Bulletin* 25, 2.

Escobar, A. (1991) 'Anthropology and the development encounter', *American Ethnologist* 18, 4: 658–82.

—— (1992) 'Planning', in Wolfgang Sachs (ed.), *The Development Dictionary: A Guide to Knowledge as Power*, Johannesburg: Witwatersrand University Press; London and New Jersey: Zed Books.

—— (1995) *Encountering Development: The Making and Unmaking of the Third World*, Princeton, NJ: Princeton University Press.

Fairhead, J. and Leach, M. (1996) *Misreading the African Landscape: Society and Ecology in a Forest-Savanna Mosaic*, Cambridge: Cambridge University Press.

Fallers, L.A. (1974) *The Social Anthropology of the Nation State*, Chicago: Aldine.

Feldman, J. and March, J. (1981) 'Information in organisations as signal and symbol', *Administrative Science Quarterly* 26, 171–86.

Ferguson, J. (1994) *The Anti-Politics Machine: 'Development', Depoliticization and Bureaucratic Power in Lesotho*, Minneapolis: University of Minnesota Press.

Gilbert, Elon, Khandelwal, Rajiv and Ballabh, Pankaj (1995) 'Monitoring GO/NGO/rural community collaboration for sustainable agricultural development in Rajasthan', *Working Paper No. 1: Methodology: Concepts and Approach*, March.

Gleik, James (1987) *Chaos: Making a New Science*, New York: Viking.

Grillo, R.D. (1997) 'Discourses of development: the view from anthropology', in R.L. Stirrat and R. Grillo (eds), *Discourse of Development: Anthropological Perspectives*, Oxford: Berg Publishers.

Hinton, R. (1995) 'Trades in different worlds: listening to refugee voices', *PLA Notes* 24, 21–8.

Hobart, M. (ed.) (1993) *An Anthropological Critique of Development: The Growth of Ignorance*, London and New York: Routledge.

Hulme, D. and Turner, M. (1990) *Sociology and Development: Theories, Policies and Practices*, Hemel Hempstead: Harvester Wheatsheaf.

Korten, David (1980) 'Community organisation and rural development: a learning process approach', *Public Administration Review* 40, 5: 480–511.

—— (1989) 'Social science in the service of social transformation', in C.C. Veneracion (ed.), *A Decade of Process Documentation Research: Reflections and Synthesis*, Quezon City: Institute of Philippine Culture, Ateneo de Manila University, pp. 5–20.

Long, Norman (1977) *An Introduction to the Sociology of Rural Development*, Tavistock Publications.

Long, N. and Long, A. (eds) (1992) *Battlefields of Knowledge: The Interlocking of Theory and Practice in Social Research and Development*, London and New York: Routledge.

Long, N. and Villareal, M. (1994) 'The interweaving of knowledge and power in development interfaces', in I. Scoones and J. Thompson (eds), *Beyond Farmer First: Rural People's Knowledge, Agricultural Research and Extension Practice*, London: Intermediate Technology.

Luhmann, N. (1969) *Legitimation durch Verfahren*, Luchterhand: Neuweid.

Marsden, D.J. and Oakley, P. (1990) *Evaluating Social Development Projects*, Oxford: Oxfam.

Marsden, D.J., Oakley, Peter and Pratt, Brian (1994) *Measuring the Process: Guidelines for Evaluating Social Development*, Oxford: Intrac.

Mosse, D. (1994) 'Authority, gender and knowledge: theoretical reflections on the practice of participatory rural appraisal', *Development and Change* 25, 3: 497–526 (earlier draft as *ODI Agricultural Administration (Research and Extension) Network Paper No. 44*).

—— (1995a) 'Local institutions and power: the history and practice of community management of tank irrigation systems in south India', in N. Nelson and S. Wright (eds), *Power and Participatory Development: Theory and Practice*, London: Intermediate Technology.

Mosse, D. (with the KRIBP team) (1995b) 'People's knowledge in project planning: the limits and social conditions of participation in planning agricultural development', *ODI Agricultural Research and Extension Network Paper No. 58*, July.

—— (1995c) 'Social analysis in participatory rural development', *PLA Notes 24*, 27–33.

Mosse, D. (1996) 'The social construction of "people's knowledge" in participatory rural development', in S. Bastian and N. Bastian (eds), *Assessing Participation: A Debate from South Asia*, New Delhi: Konark Publishers.

—— (1997a) 'Colonial and contemporary ideologies of community management: the case of tank irrigation development in south India', *Modern Asian Studies*.

—— (1997b) 'The ideology and politics of community participation: tank irrigation development in colonial and contemporary Tamil Nadu', in R.L. Stirrat and R. Grillo (eds), *Discourses of Development: Anthropological Perspectives*, Oxford: Berg Publishers.

Mouly, V.S. and Sankaran, J.K. (1995) *Organisational Ethnography: an Illustrative Application in the Study of Indian R&D Settings*, New Delhi: Sage Publications.

Nelson, N. and Wright, S. (eds) (1995) *Power and Participatory Development: Theory and Practice*, London: Intermediate Technology.

Padaki, Vijay (ed.) (1995) *Development Intervention and Programme Evaluation: Concepts and Cases*, New Delhi: Sage Publications.

Pels, P. and Nencel, L. (1991) 'Introduction: critique and the deconstruction of anthropological authority', in P. Pels and L. Nencel (eds), *Constructing Knowledge: Authority and Critique in Social Science*, London: Sage Publications.

PLA Notes: Notes on participatory learning and action (1995) 'Critical reflections from practice', No. 24, October, London: International Institute for Environment and Development.

Pottier, J. and Orone, P. (1995) 'Consensus or cover-up? The limitations of group meetings', *PLA Notes* 24, 38–42.

Rew, A. (1997) 'The donor's discourse: official social development knowledge in the 1980s', in R.L. Stirrat and R. Grillo (eds), *Discourses of Development: Anthropological Perspectives*, Oxford: Berg Publishers.

Richards, P. (1995) 'Participatory rural appraisal: a quick and dirty critique', *PLA Notes* No 24, 13–16.

Robertson, F. (1984) *People and the State: An Anthropology of Planned Development*, Cambridge: Cambridge University Press.

Sachs, Wolfgang (1992) (ed.) *The Development Dictionary: A Guide to Knowledge as Power*, Johannesburg: Witwatersrand University Press; London and New Jersey: Zed Books.

Salmen, L. (1987) *Listen to the People: Participant–Observer Evaluation of Development Projects*, New York: Oxford University Press.

Scoones, I. and Thompson, J. (eds) (1994) *Beyond Farmer First: Rural People's Knowledge, Agricultural Research and Extension Practice*, London: Intermediate Technology Publications.

Spencer, Jonathan (1989) 'Anthropology as a kind of writing', *Man* 24, 145–64.

Uphoff, Norman (1992) *Learning from Gal Oya: possibilities for participatory development and post-Newtonian social science*, Ithaca and London: Cornell University Press.

Van Maanen, J. (1979) 'Reclaiming qualitative methods for organisational research', *Administrative Science Quarterly* 24, 520–6.

Veneracion, Cynthia C. (ed.) (1989) *A Decade of Process Documentation Research: Reflections and Synthesis*, Quezon City: Institute of Philippine Culture, Ateneo de Manila University.

Volken, H., Kumar, A. and Kaithathara, S. (1982) *Learning from the Rural Poor: Shared Experiences of the Mobile Orientation and Training Team*, New Delhi: Indian Social Institute.

Wade, R. (1993) 'How to make "street level" bureaucracies work better: India and Korea', *IDS Bulletin* 23, 4.

—— (1985) 'The market for public office: why the Indian state is not better at development', *World Development* 13, 4.

Wolf, E. (1982) *Europe and the People Without History*, Berkeley: University of California Press.

Wood, G. (1985) 'The politics of development policy labelling', *Development and Change* 16, 3: 347–73.

Wright, S. (1994) *Anthropology of Organisations*, London and New York: Routledge.

Wright, S. and Nelson, Nici (1995) 'Participatory research and participant observation: two incompatible approaches', in N. Nelson and S. Wright (eds), *Power and Participatory Development: Theory and Practice*, London: Intermediate Technology.

2

PROCESS DOCUMENTATION RESEARCH AND PROCESS MONITORING

Cases and issues

David Mosse

Process monitoring and research involves approaches and methods ranging from intensive field work by full-time independent researchers (e.g. Salmen's (1987) participant-observer evaluators) to periodic reviews and reflections by agency field staff; and from open-ended ethnography-like field notes, to highly structured and selective reporting on significant events. This range is best explored through examples, and that is the purpose of this chapter. The examples start with information systems which emphasise external research and documentation (usually referred to as process documentation research, or PDR) and then move to those in which process work is more part of the stream of development initiatives itself. In reviewing different methods, a number of issues will be raised.

Process documentation research (PDR)

The term 'process documentation research' was coined at a workshop in the Philippines in 1978 as a label for research into the field-level implementation of a pilot programme to improve communal irrigation by developing effective farmer institutions for irrigation management (planning, design, construction, operation and maintenance) set in motion by the Philippines National Irrigation Agency (NIA) (de los Reyes 1989: 24). This 'community management' programme was informed by the wider policy objective of cost recovery and making the NIA self-financing. However, there was much uncertainty as to how best to promote participatory irrigation. To help develop the right approach, two pilot projects began in 1976 as 'laboratories' for learning about farmer participation and local institutional development. From 1979 full-time social scientists from the Institute of Philippine Culture took up residence in project villages (along with project

31

staff) and began detailed observation and documentation of the processes of user group formation and functioning. The objective was to use information generated by these 'process documentors' to learn from implementation experience and in this light to modify strategy and policy.

True to the 'learning process approach' adopted by the programme, the monthly process reports written by field-based documentors produced information and insights (successful approaches, problems, etc.) which were used 'not to solve the problems of the specific pilot projects, but rather to develop agency capacities to deal with problems on a program-wide basis' (Bagadion and Korten 1991). Through PDR reports there was a direct feedback of field-level experience to decision makers at different levels: to project staff, to a special NIA working group on communal irrigation, and at the national level to a committee involving a range of agency, government and donor 'stakeholders' (NIA, IPC, bureaucrats, management institutes, IRRI, and the Ford Foundation). This meant that PDR could both improve operational procedures and inform the later expansion of the programme. Indeed by 1982 there were 130 participatory projects and by 1983 all communal irrigation schemes were based on water users' associations (de los Reyes 1989: 22).

The principal feature of this PDR was the placement of specially trained and supervised researchers from outside the agency at village level who attended and observed all project activities, interviewed community and project participants (farmers, government employees, community organisers, farmers, etc.), analysed existing records (meetings minutes, project reports, financial records, etc.) and generated data on project actions and interventions, attitudes and expectations. There was one full-time documenter for each irrigation project (covering 200–300 hectares). The researchers were trained in participant observation, interviewing and narrative reporting. However, their work was not comparable to that of an anthropologist. Rather than being open-ended, observations and notes were structured by programme concerns, namely the interpretation and application of NIA policies and procedures in the field. Indeed, process documentation focused on the 'mechanisms' for developing farmer associations and enabling them to meet the management requirements of irrigation. PDR contributed to the drafting of manuals on field intervention methods (how to organise meetings, set up record-keeping systems, manage association funds, promote effective leadership or water distribution systems, etc.), curricula for training courses and in other ways building new capacities within the agency (de los Reyes 1989; Veneracion 1989b: 102). PDR did not intend to assess the development model, but to show how it could best be put into effect.

In later stages of the programme the methods were modified and rationalised. PD researchers began to work in pairs rather than singly, the field of observation became more issue-focused and carefully guided so that similar observations were made on each water users' association. Visits became less frequent (e.g. quarterly) and involved some reconstruction of events rather than direct observation (interviewing rather than participant observation), and reporting was more selective (and briefer). The shift is described as one from process documentation research to process monitoring research (PMR) (Veneracion, contribution to ODI/CDS April 1995 workshop).

The PDR, first developed with the Philippines NIA, was subsequently transferred to other sectors and regions, including social forestry initiatives in the Philippines (Borlagdan 1987), and minor (tank) irrigation in Thailand (Paranakian 1989) and Tamil Nadu (India). In the latter case,[1] village-based process documenters were integral to local site-based project teams (which also included an Institutional Organiser and a Technical Assistant) charged with the promotion of water users' associations (WUAs). These took contracts directly from the Public Works Department, organised construction work for tank systems repair and improvement, and managed tank resources. Monthly process documentation reports recorded transactions involved in negotiating tank development plans and implementation arrangements, documented activities and perceptions, conflicts and resolutions. Much of the reporting concerned the interactions of villagers with the project, its staff and other institutions (Revenue, PWD, local NGOs, etc.), and the significance of wider structures and social processes to local institution building (e.g. party political contests, leadership struggles, resistance from contractors). The reports are a valuable record of the opinions, attitudes and judgements of key project and non-project actors. The outputs of this process documentation were reviewed by project staff and by a state-level steering committee. PDR continued in the first (four) experimental projects for 5–6 years and served to influence the design of an expanded phase of the project, as well as contributing to wider state (Tamil Nadu) policy on participatory irrigation. However, process documenters were dropped from the teams in the expanded phase.

In a third PDR case, an NGO (the Aga Khan Rural Support Programme in Gujarat, India) contracted two fully independent research institutions[2] to document project initiatives in social forestry and joint forest management as well as in participatory irrigation (Shah n.d.; Parthasarathy and Iyengar, this volume). In contrast to the previous cases, PDR was an entirely separate exercise, weakly integrated into project structures and activities. Trained field observers worked for pre-defined periods (rather than continuously), reconstructing the history of implementation, recording perceptions of events, opposition and conflict resolution measures (etc.) from minutes of meetings, diaries and project files, and discussions with agency staff and village group members and non-members. They also undertook village profiles, household-level interviews and participant observation of key activities. Among other things, this PDR provided insight into the different perspectives on project objectives and activities of different actors,[3] the interaction of the project with local social structures, and the role of the NGO. Process researchers provided critical feedback to the NGO on issues such as technology choice, selective participation or the narrow base of many village organisations, but were not represented on the NGO's decision-making bodies empowered to act on this information. Parthasarathy and Iyengar(this volume) describe this retrospective evaluative approach as 'process evaluation research'. The same research agency has subsequently become involved in concurrent documentation (PDR) of the new state policy of 'participatory irrigation management' (PIM) in which irrigation management is 'turned over' from the state to water users' associations, and in this capacity the process documentation researchers now participate in a state-level working group formed to review implementation of the policy in Gujarat much as they have in Tamil Nadu and the Philippines.

Process documentation research – general issues

As these examples suggest, until now PDR has mostly been used in developing new agency capacities for promoting user groups or new management arrangements for natural resources development (e.g. participatory irrigation or forestry) in south and south-east Asia (India, Philippines, Thailand and Indonesia). The evolution of long-term strategies has justified such an intensive monitoring approach. In fact, PDR is not conceived as a monitoring device at all, but rather as a means to develop agency capabilities in community approaches prior to programme expansion. The intensity of work and the demand for the continuous presence of a trained field worker (usually supported by supervisors) limits the coverage of PDR and is compatible with exploring in depth the kinds of issues involved in local institution development in order to evolve strategy rather than to report on progress.

PDR programmes have certainly contributed to the development and understanding of processes of institution building for local resources management. They enable an analysis of styles of leadership, the shifting bargaining power of different groups and the establishment of organisation rules and systems. Equally importantly, PDR has recorded the role and influence of outsiders (project staff, bureaucrats, political leaders, etc.) in the operation of resource management groups and therefore helped overcome the fiction of autonomous self-managed local institutions, and in other ways contributed to an understanding of user organisations as dynamic and evolving political institutions (cf., Mosse, forthcoming). Finally, PDR has helped isolate key principles and issues, and contributed to wider policy on participatory resources management.

Several factors contribute to the success of PDR (measured in terms of the generation of usable data which influence policy and programme decision making). Within the Philippine NIA, factors included the strong senior management desire to learn about the community processes on which PDR reported, the existence of a senior executive (national) committee to manage the learning process, the expectation of expansion of the approach being documented, and the ability of social scientists to take on new roles (de los Reyes 1989: 38–41).

Not all PDR efforts have shared these advantages. The absence of a clear purpose (e.g. initiation of a new approach or preparation for expansion), weak management support or unclear channels for information feedback and use of data generated can result in problems or reduce the value of PDR. Some degree of resistance from agency staff and their perception that the exercise is academic, irrelevant or threatening is not uncommon (Paranakian 1989: 82; Shah, n.d.). The tension between researchers and project staff is obviously amplified where (as is often the case) some evaluative intent is perceived in the PDR (Paranakian 1989). This is part of the wider problem of legitimacy of process work mentioned in Chapter 1 and to which I return below.[4] There is also the reverse problem of too close an identification of the PD researcher with the purposes of the project and the social identity of its workers. The loss of critical distance means loss of the ability to perceive different perspectives, identify communication gaps between farmers and project staff or more generally to generate critical reflection on practice.

It was also important that PDR in the Philippines (as in Tamil Nadu) began at the very start of the project. Other PDR works such as AKRSP's which did not begin with the project found disadvantages in generating retrospective rather than concurrent data. However, while perhaps particularly valuable at the outset of a project initiative, the need for detailed analytical information is not continuous during the life of a project. Both the Philippines and Tamil Nadu participatory irrigation projects abandoned or changed process reporting when they shifted from very small-scale experimental projects (largely conceived of as 'research' exercises) towards larger programmes. There was no need for a permanent capability for PDR.

This capability is anyway difficult to establish. Cost is certainly one factor, although costs may be small in comparison to the overall project outlay. Salmen (1987) monitored projects with loan amounts of over US$9 million for about US$7,000 each. Any detailed research exercise will be costly in terms of time, and the defining feature of PDR is the recording and analysis of events *as they happen*. PDR can, therefore, only move as fast as the events which it records and analyses. Clearly, PDR is not an addition to the repertoire of rapid research methods, and cannot be compared with rapid and participatory learning methods (RRA/PRA). Having said that, there are potential trade-offs between detail/continuity and time. PDR samples events over time as well as in terms of the number of sites covered, and the sample size can be increased or decreased by monitoring for longer or shorter periods. While in the Tamil Nadu tank irrigation project 'process documenters' closely observed and recorded water users' associations in four experimental tanks over 5–6 years, in another recent study, process documenters were stationed for just one cultivation season to document the working of user associations in twenty sites in three states (IIM/IIMI 1995). Some costs can also be reduced by integrating process research more into project activities and using project staff rather than outsiders (see the KRIBP case below).

Perhaps more restrictive than time are the exceptional skills required of PDR researchers. PDR methods place considerable demands on the documenters and require a range of sophisticated skills. Researchers are required to have intimate knowledge of both the project and the village in which they work. They have to establish a critical independence from the project and its goal, good rapport with villagers and yet the ability to perceive different points of view of different actors. They need considerable skills in observation, reporting, and sufficient local and subject knowledge to enable the interpretation of events and the attribution of significance.

Unsurprisingly, it is difficult to find all these skills together. Often they have to be developed over time on-the-job. With limited skills, reporting is often restricted to descriptive and chronological accounts of events, problems, and achievements. An over-emphasis may be given to note taking and the production of reports itself as against observation, discussion and analysis. Often this reporting focuses largely on formal 'project activities' (action and events involving project personnel, etc.) and can miss (or dismiss) related local events. The problem of routine reporting of project activities can, to some extent, be addressed through the use of checklists and frameworks which prompt more in-depth enquiry. Inevitably researcher skills develop with experience.

Even where process reports are of a high quality there are problems in the *use* made of this information. Although PDR has potential as a method for policy research, in many cases this potential is under used. While the NIA 'national committee' and similar bodies ensure that relevant PDR material is available to policy makers, often it is not in readily usable form. PDR, like good ethnography (Spencer 1989: 146), is intended to clarify. But to do this an interpretative analysis is needed – to provide what Geertz (1973) called 'thick description' rather than undigested information or 'thin description'. Since PDR generates a vast amount of material, but much of it remains unused because poorly integrated, it is questionable whether the detail involved in the method is appropriate to the ends. Perhaps at a more local level this detail would be better used (for example, in transferring experience from one village to another). But this could probably be better achieved verbally. Even with more central decision making, verbal reporting is more important; 'in many instances it is not the process documentation report itself that becomes the basis of action, but rather the discussions with the process documentor or supervisor participant as a key informant' (Korten 1989: 17). Oral reporting is certainly quicker and cheaper, but it lacks the capacity that written records have for precision, developing longer-term perspectives, wider dissemination or holding parties to agreements made.

Written PDR reports have been of use as data for longer-term academic research directed towards the analysis of broader policy issues.[5] However, the interpretation of PDR material is not straightforward. The term 'documentation' attributes a spurious value-free objectivity to an activity which clearly involves active selection (judgements about relevance) and interpretation. PDR texts therefore often need to be viewed in the light of value judgements of their writers and the representations (or misrepresentations) of rural society which they reveal. As such PDR can indeed provide material for critical discourse analysis of another kind.

Theoretical difficulties aside, the practical efforts in setting up, supporting and learning from PDR mean that its use (and justification) is likely to be restricted to experimental and innovative pilot projects (and where, as in NIA, there is strong institutional motivation to learn), where implementation contexts are unusually complex and long drawn out (involving, for example, new forms of collaborative work, or collective action), or where replication is likely (Veneracion 1989b: 88). PDR is not a generally applicable approach and is not appropriate in the absence of these sorts of exceptional complexity, innovation or expansion and replication.

But, as the Philippines case illustrates, methods can be adapted. A shift away from long-term participant observation and analytical description towards shorter, issue-focused research which incorporates rapid and participatory tools (RRA/PRA) may have wider application within projects. Other projects have moved away from the idea of PDR as a research exercise separate from and running parallel to programme planning and implementation. An alternative approach, then, is to integrate process 'research' into a project and its own systems of information management and decision making; to develop and make a virtue out of that blurred area between research and action. Here we shade from process documentation research into *process monitoring*.

Process monitoring within individual projects

In the following two cases process information is, like NIA, used for adaptive change and the development of new capacities for participatory development in agencies, but its production is significantly rationalised and undertaken by the project itself, rather than by a specialist agency.

The Kribhco Indo-British Rainfed Farming Project (KRIBP) is an ODA-funded project located in tribal western India (see Jones *et al.* 1994). The objective of the project is to develop (and implement) a participatory *approach* to farming systems development within a large and bureaucratic parastatal organisation unfamiliar with community-based approaches. Organisational learning and the incorporation of approaches used success-

37

fully within NGOs were central to the project's task. The documentation of project processes and critical reflection on initiatives were therefore seen as crucial to the evolution of project strategy, field methods and procedures. Unlike the PDR cases, however, this did not involve the use of specialist external 'process monitors' but rather involved information feedback from different actors within the project itself. Like many donor-funded projects, KRIBP involves a number of different agencies with different agendas – the project unit (KRIBP), the agency of which it is a part (KRIBHCO), consultants (from the University of Wales), donors (ODA) and government. Within each of these there are distinct interests and concerns defined by position (e.g. field workers, project management, senior agency executives), discipline (agronomists, foresters, soil engineers, anthropologists) and development orientation (overlapping concerns with production, environment, income, poverty, gender equity). The 'project' itself was a negotiated, but frequently unstable, operating consensus among these different players and perspectives, and the demands of villagers themselves. 'Process monitoring' often had to do with developing, negotiating, modifying and supporting some degree of workable consensus and implementing the 'project's approach', while acknowledging the existence of conflicting interests and concerns.

Much of the most immediate non-routine information feedback resulted from the interaction between project staff and external consultants, either technical (relating to specific programmes) or more general (concerning planning, institutions, or gender). These periodic and collaborative reviews involve short-term field work, verbal reporting, review workshops, and the analysis of monitoring information. They served to review trends and impacts, analyse failure, test and significantly adapt project strategies in the light of local information. This type of 'process' monitoring is quite common in externally assisted projects. The difference is that here more attention was given to the documentation, analysis and wider dissemination of the course of project learning in key areas in terms of the broader policy and methodological issues raised (e.g. Mosse 1994, 1995b, 1996).[6]

There were also attempts to generate more open-ended information on field-level social contexts and the processes associated with project interventions to feed into this 'process monitoring'. These included field worker diaries, 'process files' (both introduced to try and emphasise separation from the routine log of activities), notes on meetings and activities and case-study 'stories'. However, none of these were very successful. Field staff face many pressures and narrative reporting is neither a priority nor something for which many have the skills, aptitude or interest. Nonetheless, field workers are generally good observers and analysts of the social situations in which they operate. Indeed they have to be in order to carry out work and pursue project objectives in local communities. This requires a sophisticated understanding of local power relations, spheres of influence, social tensions, and areas of support and opportunity. Value was seen in making this tacit analysis explicit: to recognise and give value to field worker skills, to encourage sharing and wider organisational learning, and to provide critical feedback on project strategies.

This was attempted in the project through *ad hoc* local-level (village cluster) semi-structured review meetings/workshops involving project staff. These used observations on project activities (patterns of participation, withdrawal, collapse, obstruction, cooption or control of activities) to provide insights into the local power relations and networks of influence that would influence project work locally. These were initially facilitated by visiting anthropologists and gender consultants who documented and provided critical commentary on the project's evolving strategy on participatory planning and institutional development (e.g. Mosse *et al.* 1995a). Following this there were further efforts to institute more regular local-level 'review' workshops employing a set of guidelines for the critical review of events in the project villages and facilitated and documented by an outsider social scientist. For several reasons it proved impossible to routinise process monitoring into the project in this way. First, there were problems in identifying facilitators. Second, pressures from routine work and the demands of routine monitoring reports directed interest away from process monitoring. Third, there was a lack of clarity about the purpose of process monitoring and an absence of clear feedback channels to management for review and decision making. Partly for this reason, fourth, there was limited interest from the potential users, whether donors or agency executives, of project management. Process documentation has continued but has largely fed into more analytical work for wider dissemination

In a second project case, Davies (this volume) has evolved a response to the triple challenge of rationalising information production, focusing on the concerns of project actors and yet retaining inductive open-endedness. Davies describes an innovative approach to organisational learning providing an experimental means to monitor change and impact without using conventional indicators. This he developed with an NGO, the Christian Commission for Development in Bangladesh (CCDB). The crux of the system is the deliberate selection and recording of a small number of change stories of high 'significance'. The first task is to select the key 'domains of change' to be monitored, which in this case are changes: (1) in people's lives, (2) in people's participation, (3) in sustainability, and (4) in any other area. Significant change events are then selected and reported by different groups at different levels – by village group members, field staff, head office staff and by groups representing donors. Each group devises its own basis and methods for selecting the events to report, either from direct experience or from field reports.

At each level, a descriptive (verifiable) account is accompanied by an explanatory account stating *why* particular changes are considered important, or what difference they have made or will make. In this way the basis and methods of selection are made explicit and value judgements are brought into the public domain. The approach consciously departs from the process documentation myth of 'uncensored field experience' or 'unedited truth' (Korten 1989: 14; Veneracion 1989b: 98). The results of reporting and selection are fed back from head office to field teams. The most significant accounts are subjected to factual verification and quantification (e.g. what was the frequency of significant incidents in other locations?). Further analysis identifies the proportion of local groups (and field staff) contributing to 'significant changes' and the correlation between this and other features (e.g. savings levels, group size or gender composition).

The method provides a workable monitoring system. It is open-ended and accepts and expresses diversity. The inter-subjectivity of the system allows explicit recognition of different perspectives and the role, subjectivity and value concerns of the observers. Agreement on meanings is a *result* and not, as usually, a precondition for the monitoring system. Summarising experience by progressive selection, rather than inclusion (enumeration, averaging) differentiates rather than homogenises and retains the true diversity of reported experience. The method is inductive and open-ended rather than deductive and closed, and unlike the use of conventional indicators, monitoring events are identified and recorded after they have occurred rather than anticipated (deduced from design) at the outset. Information is produced, interpreted (and used) (i.e. processed) by persons close to the events themselves rather than centralised research units. In consequence monitoring 'stories' retain contextual information as well as providing a dynamic medium in which to record changing perceptions and events.

In contrast to the experience of many carefully designed monitoring systems (including process monitoring in KRIBP) the CCDB system was accepted, continued and extended by the project management. Importantly, the system has encouraged diversity in reporting and avoided the tendency towards a 'steady state' and has provided information widely used in the organisation (e.g. for promotional material) (Davies, this volume). However, while clearly highly responsive to the needs of the organisation, the process monitoring method, Davies suggests, did not particularly encourage critical or analytical comment. This was explicitly the focus of the first of the next paired cases.

Process monitoring in inter-agency settings

The following two cases shift from individual projects to more complex inter-agency settings. They illustrate contrasting understandings of process monitoring. In the first case process monitoring is 'an analysis of how inter-agency partnerships work', while in the second it becomes 'a strategic involvement in inter-agency collaboration'. The contrast is between process monitoring as 'institutional ethnography' (Lewis, this volume) and as a set of facilitating 'communication services' (Farrington, Gilbert and Khandelwal, this volume). There are corresponding contrasts in the perception of process monitoring by the agencies concerned which illustrate more general problems in the organisational politics of monitoring and information control. Unsurprisingly the analytical-critical approach proves less compatible with organisational interests than the 'communication services' approach and for this reason has run into problems of legitimacy. Let me briefly set out these contrasting cases beginning with Lewis's 'institutional ethnography' of Bangladesh fisheries institutions.

The USAID-funded International Centre for Living Aquatic Resources Management (ICLARM) project for 'Technology Transfer and Feedback through NGOs' focused on the development of low-cost technologies for poorer Bangladeshi farmers. It involved a complex framework of institutional partnerships between government agencies and NGOs which sought to strengthen the capacities of both types of agency through collaborative links drawing on complementary competencies in technical knowledge and community approaches. An ODI research project was set up to examine the nature and functioning of these partnerships, their history and performance (Lewis, this volume). Methods included tightly structured problem solving workshops followed up by semi-structured interviews with participants. Tape and video were also used to record discussions, and the content analysis of these helped to outline a 'benchmark' of assumptions, attitudes and experiences against which progress could be reviewed, problems acknowledged and solutions evolved.

The practical result of this monitoring was to make a 'top down' institutional culture more adaptive and to encourage, for example, the adaptation of a technology 'packaged' to better suit different agro-ecological priorities. The process monitoring itself helped improve inter-agency (GO–NGO) communication, strengthened the hand of NGOs in making inputs into government fisheries policy, and this, in turn, re-kindled some defunct NGO networks.

In addition to these constructive outcomes, the research also shed light on the different motives agencies had for collaboration and the wider significance of, and constraints to, inter-agency partnerships and linkages in the context of competition for scarce resources. As Lewis, speaking of the two principal technical agencies, puts it, 'ICLARM and FRI need each other far more for the individual institutional survival of each agency than the average low income farm household in Bangladesh needs new technology for aquaculture. FRI clearly needs a donor patron' (Lewis, this volume). These institutional needs perpetuated mis-specifications of the development problem (in this case portraying technology constraints rather than socio-economic issues of resources access and tenure as the principal bottlenecks to aquaculture development). Moreover, since they were built upon the need to secure external resources, these partnerships and the organisational changes they were intended to produce proved to be short lived beyond the life of the project.

In drawing attention to the interested construction of partnership, distorted linkages, or limited learning, this 'institutional ethnography' hit upon sensitivities. As outsider-initiated research it was vulnerable to manipulation and resistance and its limited 'legitimacy' to key stakeholders eventually led to a premature end to the project.

The tensions and partial failure of the ICLARM process research project provided important analytical insights into the workings of an inter-agency project, in particular the way in which organisational interests generate a particular discourse of 'partnership' and construct development problems and solutions (Lewis, this volume). The experience also pointed to the need for sensitivity in process monitoring and awareness of the risks inherent in process research in donor-funded programmes. The single-minded pursuit of particular interpretations of events irrespective of the interests of other actors had indeed contributed to the termination of the ICLARM process research project. The need for flexibility and sensitivity to different perspectives and interests is explicitly addressed in the next case, also concerned with NGO-government agency collaboration and undertaken by ODI. In this instance, however, there was no link to external (donor) funding, and no bounded project or programme. These features, and the insistence on 'being guided by what organisations perceived to be useful in documentation, analysis and communication', produced a very different kind of 'process monitoring'.

The focus was on interactions between NGOs and government agencies in agricultural research and extension in a district (Udaipur) in south Rajasthan (India). As Farrington, Gilbert and Khandelwal's account (this volume) makes clear, 'process monitoring' was initially viewed as having two purposes: the first, to assess 'the impacts of collaboration on organisational performance and on rural communities' and the second, to provide the means to promote inter-agency understanding and collaboration. In the event the 'researchers' felt that the analytical monitoring and impact assessment objectives were unrealistic, and, more importantly, unacceptable to participating agencies. Consequently there was a decisive shift from the analysis *of* collaboration towards a strategic involvement *in* collaboration. In roles described as 'advocacy' and 'nurturing', a range of documentation methods (village studies, Working Papers, newsletters, published letters or records of agreements) were used to promote networking, facilitate negotiation, enhance leverage, increase accountability and exert policy influence. These documentary methods served to give marginal actors increased voice, and to publicise and legitimise progressive changes (e.g. increased responsiveness of government programmes to NGO and community inputs). The role of outsider 'researchers' here was not observation and analysis but support, for example, in gaining access to senior officials. Process monitoring amounted to providing a 'communication service' to address local concerns and to resolve immediate difficulties.

A rather similar responsiveness to the documentation needs of villagers was involved in the documentation of a housing project among a Dalit group in Gujarat (Virami 1995). This documentation was not set up to meet the need of the NGO or officials but of the community for whom, as Virami puts it, 'housing represented a monument to a long and tortuous struggle over many years . . . . The history of the struggle over the last decade had to be written afresh, it had to be written as a record of collective triumph' (Virami 1995: 222). This particular housing project was rooted in caste conflict over land and its significance was as struggle against social subordination. The language of documentation therefore had to use the 'emotional vocabulary' of a people engaged in the redefinition of their social identity (the feelings of fear, indignity, anticipation, pride etc.) (Virami 1995).

This process documentation was a means to validate a process of social change. Process work in Udaipur has also sought to validate collaborative work. It was also concerned to resolve complex communication problems. The themes of 'validation' and 'resolution' are given explicit focus in Rew and Brustinow's (this volume) work on policy reform in Russian farming and Indian forestry. Tracing the monitoring of ODA-assisted policy reform initiatives, Rew and Brustinow (like Lewis, this volume) draw attention to the diverse institutional agendas behind official agreements on policy reform. As mentioned already, they see policy innovation as generating ambiguous procedures, and uncertainty about benefits. Monitoring has the important role of validating (confirming, recognising outcomes) objectives and resolving differences and ambiguities, especially in face of opposition to change and reform. As Rew and Brustinow put it:

> The monitoring system may be less a means – either 'quantitative' or 'qualitative' – to manage the ultimate end of 'impacts' or 'transactions'. Rather, the monitoring system must play a major part in establishing the framework of discussions which negotiate the common meanings and resolutions which in turn allow the reinterpretation of positions and the derivation of reassurances about the negotiation and processing of benefit and loss.
>
> (Rew and Brustinow, this volume: p. 187)

As in the ODI-Udaipur case, process monitoring is viewed as 'embedded within, not as something beyond or outside, the institutions which guide the policy reform process; it is part of an institutional and policy learning process rather than a set of tools for project assessment' (Rew and Brustinow, this volume).

As Rew and Brustinow stress, within process-type projects, 'resolvents' provided through monitoring are necessary to re-focus controversy and negotiate consensus and ends. It is significant that in one of the two descriptive cases (the Russian case), process monitoring succeeded in positioning an 'institutional *resolvent*' while in the other case (Indian forestry, as in the Bangladesh fisheries case) it failed to acquire the necessary legitimacy and validation to achieve this.

Critical concerns

The comparatively limited experience of process-oriented research and monitoring already raises a number of critical issues, some of which were addressed in connection with 'process documentation research' (PDR) above. The issues of purpose, legitimacy, cost and coverage, and written versus oral reporting, for example, are relevant to the full spectrum of process approaches. A number of further points need comment.

One set of issues concerns the methods themselves. First there is the question of *who* should be involved in information production – insiders or outsiders. The different process monitoring and research experiences discussed here give different emphasis to 'insiders' and 'outsiders', to community members, project staff or specialists. On the one hand, independence from routine project affairs is an important element in all process monitoring and documentation work and necessary to initiate new streams of information, new methods and to develop reflective insights. There are also practical reasons for favouring external researchers. Constraints of time and interest mean that project staff often are not able or inclined to undertake process documentation work or even to analyse routine monitoring data sufficiently. The benefits of using outsiders, however, depends upon these researchers having a sufficiently long-term association with a project to understand its context.

On the other hand, undertaking process monitoring from within the agency has the advantage of increasing the relevance and acceptability of the exercise as well as improving the chances of feedback into decision making and agency learning. Moreover, as Davies (this volume) shows, it is the very engagement in events which generates the perceptions and judgements which are a central part of the process being understood. 'Insiders' also offer the advantage of continuity and a better understanding of the area and context. Arguably, the closer the process monitoring is to routine agency work and monitoring the more likely it is to be sustained; but equally the less likely it is to add something new. Information routinely collected is likely itself to become routine. It is therefore necessary to find ways of distinguishing process from routine monitoring. Under some management regimes critical observation and reflection can be introduced by 'role shifting' and allowing field practitioners to become observer-documentors for a time. In addition, process work by 'insiders' may need to occur irregularly and involve reflection over a sufficiently long period to allow programme staff to notice patterns (of participation, of implementation failure, etc.).

These points notwithstanding, in many multi-stakeholder projects involving funding agencies, line departments, NGOs (etc.), the distinction between 'insiders' and 'outsiders' may not be very helpful. The difference between insiders and outsiders may also be far from clear to community members themselves. All process documentors and monitors therefore face the difficult task of establishing a clear and legitimate identity for themselves and their activity.

A second and related issue concerns the relative merits of participatory and participant observation methods. The former are likely to increase the legitimacy and accountability of process monitoring and research (both 'upward' and 'downward' accountability). But a question remains as to how much *downward accountability* to programme beneficiaries (or 'primary stakeholders') there is, particularly where there is an emphasis (a) on written information which greatly reduces accessibility of data to those whom they describe, and (b) on feeding-up information for strategic or policy change (Montgomery, workshop paper)? How can the tendency for 'data to be appropriated as it flows upwards' (Montgomery, workshop paper) be overcome? Is it not inevitable that the perspectives of certain stakeholders will be given privilege (even where different and competing views are recognised)? Of course these problems face all social research and are not peculiar to process work, and in some ways they can be better addressed in process monitoring, for example, through information feedback and review which give results a wider legitimacy (Bebbington, workshop paper). Nonetheless, few involved in process monitoring would underestimate the difficulty of producing an agreed text.

Participatory methods may help to ensure legitimacy and accountability, but, as my comments on PRA in Chapter 1 suggest, they may also conceal different or conflicting perspectives in the interest of retaining official views and generating planning or monitoring consensus. It is largely through 'ethnographic' (rather than participatory) approaches that critical accounts of core concepts and metaphors, prevailing consensus models or underlying organisational objectives and interests have been generated. Lewis's (this volume) 'institutional ethnography' shows that under closer observation inter-agency 'collaboration' in Bangladesh fisheries is not all that it seems, and its practice often does not produce the promised benefits. A closer critical observation of the practice of 'participation' is also well overdue. Information generation and use within participatory projects usually departs considerably from the populist ideal of 'knowledge from the people for the people', and may often be part of a project's exercise of power, constraining as much as enabling self-determined change, or having the effect of advancing organisational more than community interests (Mosse 1996).[7]

Critical analysis is necessary, but it is unlikely to be achieved through participatory methods geared towards consensus building. Neither the analysis of intricate patterns of organisational competition in Bangladesh, nor the 'social maps' of village factions, alliances and conflicts underlying local institutional development in KRIBP and other projects, could or would have been generated by the participants themselves. Certain kinds of knowledge cannot (or should not) be openly stated, shared and agreed in the public domain. The current tendency to polarise 'extractive' and 'participatory' research modes is therefore unhelpful in that it overlooks the fact that certain types of knowledge in development are necessarily analytical and external to participatory action.

A third set of issues concern the coverage of process work. It was mentioned in connection with PDR that there could be trade-off between coverage and intensity. Given the nature of process material, conventional sampling criteria have little relevance. Process research requires both the local intensity to understand the detail of development processes, and the breadth to capture the 'chance events' relevant to project impacts. There is a need to engage in detailed participant observation and to scan a wider area for significant happenings within and outside the project geographical and subject areas. In all cases, selection and the attribution of significance are a critical part of process documentation and monitoring, one about which, as Davies points out, researchers need to be more explicit.

A fourth and recurring issue concerns not the production of process information but its *use* within development agencies, and in particular its capacity to contribute to institutional change. In Chapter 1 it was pointed out that there is no necessary link between the availability of information and learning or improved decision making. The experience of process documentation and research shows that process information is only likely to be used where: (a) there is some clear purpose to its collection (and a desire on the part of senior management to use process information as a source for learning and a desire among other staff to produce it); (b) there are clear channels to feed back findings to decision makers; and (c) the raw data are sufficiently and sensitively interpreted. There are several instances where this has not been the case and where, as a result, process monitoring has not gained the support that it needed from management. Experience suggests that organisations are at different stages and more or less open to using process information for internal learning. Process monitoring and documentation is most likely to contribute to learning where there is some external environment encouraging change, such as a policy reform requiring an agency to undertake new tasks or to enter into new collaborations.

The significance of these different methodological questions depends upon the type of process monitoring/research undertaken. Despite the diversity of approaches and irrespective of the particular methods used, process work tends to be oriented in one of two directions. On the one hand, it is oriented towards the critical analysis of experience (whether of community-based development or inter-agency collaboration), research and learning; often in preparation for organisational change. Such an orientation may be directed towards the production of 'institutional ethnography', or an under-standing of the 'architecture' of a project, including the structure of its interests and motives (e.g. Lewis, this volume). Process monitoring concerns here tend to be defined outside of the programme where the ultimate consumers of the information are located.

On the other hand, other types of process monitoring/research are firmly embedded in the institutions and processes they reflect on. They are 'part of an institutional and policy learning process rather than a set of tools for project assessment' (Rew and Brustinow, this volume). The roles involved here may be described in terms of advocacy or nurturing, facilitating, negoti-ation, networking and so forth. They are oriented towards resolving problems, addressing local concerns and responding flexibly to demands (Alsop, Rew and Brustinow, Farrington, Gilbert and Khandelwal, this volume).[8] Such strategies require a willingness to allow agencies to set the agenda, to abandon certain analytical research objectives, and to respond to agency needs for problem solving or the production of promotional mate-rial.

At one level these are complementary approaches – effective 'resolution' depends upon a good analysis of conflicts and controversies. But while not entirely incompatible, these two orientations pull in different directions. As mentioned, the first tends to be oriented outwards to the concerns of donors, programme expansion, or policy reform; the second towards the working out of consensus, collaboration, or change within the programme setting. In the shape of process documentation research (PDR), process-oriented work began in the first mode (although in the Philippines with the benefit of strong agency backing). As the cases reviewed in this chapter show, it has subsequently diversified into the second mode. Some process monitoring projects themselves began with a clear analytical and research agenda which was later abandoned in favour of a more facilitative role (e.g. Farrington, Gilbert and Khandelwal, this volume).

There are several reasons for these shifts, including the expansion of the concept of process monitoring and research itself in response to new needs. Several factors, however, relate to the inherent problems in process moni-toring and research discussed at the end of Chapter 1, which have to do with its legitimacy (or rather the lack of it) within development organisations. The issue of legitimacy or the acceptability of process work is complex, but it does point inescapably to the fact that information in organisations is never viewed simply as a 'public good'. Information does not exist indepen-dently of sets of interests and relations of power and control. All organisations control information flows and information is deliberately restricted. In its attempt to interfere with existing information flows within organisations (to bypass filters etc.), process monitoring and documentation may face fundamental difficulties.

Very few management systems actually encourage and even fewer reward critical observation. These are more likely to be seen as divisive and threatening to the 'hierarchy of command'. Career advancement is more often premised upon uncritical implementation and the meeting of targets (stated or unstated). On the other hand, non-hierarchical NGO or 'people's organisation' cultures eschewing bureaucratic procedures may perceive process monitoring or documentation as an undesirable, externally imposed bureaucratic evaluation, or at worst (in situations where radical action takes underground and militant forms) as an insidious form of state infiltration or surveillance (Gerard Clarke, workshop contribution). In both cases, process research may be viewed as time-consuming sophistry and irrelevant to the achievement of outputs and wasteful of staff time and resources. At worst it will be strongly resisted for its potential to draw attention to rent seeking and other unofficial transactions (e.g. NGO payment of 'rents' or 'revolutionary taxes' to the Communist Party, or the 'leakage' of project funding to underground organisations in the Philippines in the mid-1990s, Clarke, workshop contribution).

Furthermore, in most development agencies there are strong organisational imperatives to report success. This is especially so in externally supported and evaluated programmes (NGO or government) where critical feedback may be seen as having resource implications. Documentation is most acceptable where it contributes to and serves promotional purposes (case material for reports, publications etc.). Such organisational needs may be served by dissemination which emphasises the demonstration of success, formulates models, raises project profile and encourages new high profile and 'politically' rewarding training, advisory or coordination roles. Certain types of uncritical documentation and dissemination may, however, actually paralyse organisational learning.

The analysis and documentation of the untidy business of practice, exploration, difficulties, tensions and unresolved problems rarely find management support. Process information indicates a gap between intention and action, demonstrates the weakness of prevailing consensus models, or points out contradictions. This is not overly welcome in the wider context of an institutionally grounded need for simple models, manageable worlds, usable fictions or sellable products. Beyond the immediate confines of a project and its own learning cycle, process documentation, if critical of given methods or approaches, is likely to be branded as serving marginal research interests, or as undermining innovation, or weakening positions contending for influence in national or international policy arenas: 'please don't complicate this, we've only just got it onto the agenda'.

Almost all process monitoring and research which has attempted an independent analysis of social and institutional relations or the limits to learning in projects or programmes, sooner or later has run into problems. Process monitoring independent of immediate management control quickly loses legitimacy and becomes viewed as threatening interference.[9] In several cases this has ultimately led to the closure of programmes of process monitoring (e.g. Lewis, this volume). The loss of legitimacy is expressed in various ways: process work is undermined, for example, when researchers face non-cooperation; it may be neutralised by lowering the status of the process monitors, isolating them from decision making, questioning their methods (which may fail to meet bureaucratically defined notions of 'valid data'), denying them access to meetings or documentation, or carefully circumscribing the areas in which they can work.

To some extent, the realisation that information is not an objectively valid public good, and therefore that process documentation work is fraught with problems, has encouraged innovative types of process monitoring which work *within* existing authoritative domains and attempt to shape and mould existing flows of information rather than cutting across them. Several contributions to this volume are in this vein (Davies; Farrington, Gilbert and Khandelwal; Rew and Brustinow). They suggest that process monitoring *can* provide the tools necessary to resolve conflict, and to deal with uncertainty without resort to simplistic models. While the problems in managing information in programme contexts are real, examples from India, Bangladesh and Russia demonstrate that process monitoring can acquire and retain a legitimacy which allows for an active engagement in the complexities of social relationships and the negotiation of approaches, resource inputs, procedures, activities, behavioural orientations and meanings. Iterative planning and implementation demands 'resolutory' roles which build consensus based on better understanding of different as well as shared interests.

Undoubtedly, these roles require considerable sensitivity. There is always the possibility that knowledge of 'game plans' will reduce the chances of arriving at workable consensus between different stakeholders. Consensus building involves information which is necessarily reserved, unstated, coded and otherwise circumscribed by rules of tact, discretion and diplomacy. If they bring conflicts, mismatched aims and objectives among different stakeholders or unresolved policy contradictions to light in inappropriate ways, process monitoring roles may even *reduce* the room for compromise or workable consensus. In such cases, the explicitness and scrutiny involved place additional pressures on new collaborative links or delicate negotiations, may contribute to their failure, or reduce the chances of workable consensus for different stakeholders.

There seem, however, to be grounds for optimism that, armed with a sufficiently sophisticated understanding of the politics of information flows in organisations, process monitoring can contribute constructive resolving mechanisms to, *inter alia*, planning, policy reform and inter-agency collabo-

49

ration even in difficult institutional environments. The need for such mechanisms is clear from their absence in currently popular devices for consensus building. For example, logical framework analysis is based precisely on the isolation of causal links (e.g. input–output–impact) from the institutional context. It provides an instrument for generating public consensus, and is able to do so to the extent that it identifies a common ground of agreement *and* relegates differences (underpinned by a range of political and institutional objectives) over which agreement may be difficult under the label of 'assumptions' – i.e. influences beyond management control. Maybe the fictional notion of simple causal links between designed inputs and outputs is absolutely necessary to the consensus building which moves planning forwards, but it only removes from visibility the agendas to which arguments over design relate.

The positive effect of abandoning external research perspectives and working within existing systems is, therefore, enhanced power to advance development initiatives, to create the necessary consensus, resolve differences and validate progressive change. But there are costs too. The removal of critical reflection may allow the perpetuation of mis-conceived models, may foster self-serving institutional collaboration or contribute to covering over the gaps between intention and action.

Finally, process monitoring, especially in inter-agency situations, is likely to be restricted to themes or areas of operation which are uncontentious (e.g. agricultural technology). It will have little to contribute in areas where real conflicts of interest rule out progressive resolutions (e.g. conflicts over land, water or forest resources). Here conventional strategies of advocacy, lobbying and political conflict are likely to dominate. Where interests are *least* entrenched or polarised, information has more of the characteristics of a 'public good'; where interests are deeply opposed, information production and use will be strongly linked to individual and collective political strategies. Process monitoring (and documentation), perhaps, has a definable niche within a spectrum of interest-information links, occupying that place where interests are least polarised, where development outcomes have positive sum characteristics and where information is more like a public good separable from private interests.

Notes

1 The institutional development and process documentation parts of a large EEC-funded Public Works Department tank development programme were undertaken in a few experimental tanks funded by the Ford Foundation and implemented by the Centre for Water Resources (Anna University Madras).

2 The Institute of Rural Management, Anand (IRMA), and the Gujarat Institute for Development Research (GIDR), Ahmedabad.

3 Parthasarathy and Iyengar (this volume) report, for example, that despite the environmental and production enhancement objectives of the NGO, a tank

rehabilitation project was largely perceived by villagers as an employment generation drought-relief measure (cf., Mosse 1995b for a parallel case).

4   Legitimacy problems exist in relation to community members as well as agency staff, especially where researchers lack status and are not taken seriously because unclearly related to 'the project'.

5   The existence of process documentation reports over 5 years (1988–94/5) provide many opportunities for the analysis of project processes at village level: *inter alia* the intersection of indigenous water management systems and development institutions; the articulation of local social relations (relations of caste, dependence, gender) and project activities; the influence of external agents and resources on local social relations and the kind of bargaining/negotiation involved. The analysis of monthly reports shows the way in which tank irrigation systems (tanks and their supporting institutions) in rural Tamil Nadu are not only economic resources shaped by economic interests, but also serve local political purposes, express political position and caste status as well as providing opportunities to challenge existing authority, canvass political support to articulate factional affiliation or organize caste protest, and a clearer understanding of the role of project actors in local public action helps to raise appropriate questions about the sustainability of externally promoted participatory development (Mosse 1997).

6   For this purpose a KRIBP Working Paper series was initiated.

7   People's knowledge about local livelihoods, for example, may more often be used by project staff to bargain with villagers, to negotiate compromise between short-term and long-term perspectives, as a basis for argument, to challenge locals' claims on the project, to reject as well as accept villager proposals, to negotiate subsidy levels, to allocate labour benefits, or to identify the limits of local capacity (e.g. in management or cooperation).

8   The choice of perspective may relate to personal experience. After several frustrating years of trying to nurture and enable participatory practice, I began to reflect critically on the constraints to participation, and more broadly on the politics of knowledge production and use in a participatory project.

9   The potential challenge which process documentation and monitoring (of some types) presents to bureaucratic programmes, provides a sharp contrast with PRA (participatory rural appraisal) which has been effectively incorporated and bureaucratised within large organisations. This, of course, presents major problems of its own (Mosse 1994, 1995b).

References

Bagadion, B.U. and Korten, F.F. (1991) 'Developing irrigators' organisations: a learning process approach', in M.M. Cernea (ed.), *Putting People First: Sociological Variables in Rural Development*, (2nd edn), World Bank, Washington: Oxford University Press.

Borlagdan, S.B. (1987) *Working with People in the Uplands: the Bululokaw Social Forestry Experience*, Quezon City, Philippines: Institute of Philippine Culture.

de los Reyes, R.P. (1989) 'Development of process documentation research', in C.C. Veneracion (ed.), *A Decade of Process Documentation Research: Reflections and Synthesis*, Quezon City: Institute of Philippine Culture, Ateneo de Manila University, pp. 21–44.

Geertz, C. (1973) 'Thick description: towards an interpretive theory of culture', in *The Interpretation of Cultures: Selected Essays by Clifford Geertz*, New York: Basic Books, pp. 3–30.

IIM/IIMI (1995) Preview paper for workshop on irrigation management transfer in India, Indian Institute of Management, Ahmedabad/International Irrigation Management Institute, Colombo, 11–13 December 1995.

Jones, S., Khare, J.N., Mosse, D., Smith, P., Sodhi, P.S. and Witcombe, J. (1994) 'The Kribhco Indo-British Rainfed Farming Project: Issues in the planning and implementation of participatory natural resource development', *KRIBP Working paper* No. 1, Centre for Development Studies, University of Wales, Swansea.

Korten, D. (1989) 'Social science in the service of social transformation', in C.C. Veneracion (ed.), *A Decade of Process Documentation Research: Reflections and Synthesis*, Quezon City: Institute of Philippine Culture, Ateneo de Manila University, pp. 5–20.

Mosse, D. (1994) 'Authority, gender and knowledge: theoretical reflections on the practice of participatory rural appraisal', *Development and Change* 25, 3: 497–526 (earlier draft as *ODI Agricultural Administration (Research and Extension) Network Paper* No. 44).

Mosse, D., Ekande, T., Sodhi, P., Jones, S., Mehta, M. and Moitra, U. (1995a) 'Approaches to participatory planning: a review of the KRIBP experience', *KRIBP Working Paper* No. 5, Centre for Development Studies, University of Wales, Swansea.

Mosse, D. (with the KRIBP team) (1995b) 'People's knowledge in project planning: the limits and social conditions of participation in planning agricultural development', *ODI Agricultural Research and Extension Network Paper* No. 58, July.

Mosse, D. (1996) 'The social construction of "people's knowledge" in participatory rural development', in S. Bastian and N. Bastian (eds), *Assessing Participation: A Debate from South Asia*, New Delhi: Konark Publishers.

—— (1997) 'The ideology and politics of community participation: tank irrigation development in colonial and contemporary Tamil Nadu', in R.L. Stirrat and R. Grillo (eds), *Discourse of Development: Anthropological Perspectives*, Oxford: Berg Publishers.

—— (forthcoming) 'Village institutions, resources and power: The ideology and politics of community management in tank irrigation development in south India', *Papers in International Development* (Centre for Development Studies, University of Wales). (Paper presented at International Conference on The Political Economy of Water in South Asia, Joint Committee on South Asia Social Science Research Council, American Council of Learned Societies and Madras Institute of Development Studies, 5–8 January 1995.)

Paranakian, Kanda (1989) 'Process documentation research on medium scale irrigation development', in C.C. Veneracion (ed.), *A Decade of Process Documentation Research: Reflections and Synthesis*, Quezon City: Institute of Philippine Culture, Ateneo de Manila University, pp. 65–86.

Salmen, L. (1987) *Listen To The People: Participant–Observer Evaluation Of Development Projects*, New York: Oxford University Press.

Shah, Amita (n.d.) 'Process documentation research: some reflections on methodology', MS, Gujarat Institute of Development Research, Ahmedabad.

Spencer, Jonathan (1989) 'Anthropology as a kind of writing', *Man* 24, 145–64.

Veneracion, Cynthia C. (ed.) (1989a) *A Decade of Process Documentation Research: Reflections and Synthesis*, Quezon City: Institute of Philippine Culture, Ateneo de Manila University.

—— (1989b) 'Nature and uses of process documentation research', in *A Decade of Process Documentation Research: Reflections and Synthesis*, Quezon City: Institute of Philippine Culture, Ateneo de Manila University, pp. 87–114.

Virami, Sandeep (1995) 'Role of the facilitator', in Vijay Padaki (ed.), *Development Intervention and Programme Evaluation: Concepts and Cases*, New Delhi: Sage Publications. (Excerpts from *A Struggle for Housing: A Statement by the Vankars of Goldana*, facilitated by Sandeep Virami, Ahmedabad: Janvikas and Vikas Centre for Development, 1989.)

Part 1

PROCESS MONITORING AND IMPACT ASSESSMENT IN DEVELOPMENT PROJECTS

3

PARTICIPATORY WATER RESOURCES DEVELOPMENT IN WESTERN INDIA

Influencing policy and practice through process documentation research

R. Parthasarathy and Sudarshan Iyengar

Introduction

Throughout the world state-managed irrigation systems are in serious disrepair and irrigation administrations are facing deepening financial crisis. An important response to this is the policy of participatory irrigation management (PIM). PIM aims to improve the performance and financial viability of irrigation structures through systems of consultative planning, cost recovery and the turning over of operations and maintenance to local water users themselves. To achieve this, considerable institutional innovation is needed at the local level in order to promote farmer participation in irrigation management. Furthermore, it is repeatedly found that the effectiveness of PIM depends upon an understanding of local social contexts and on having good information feedback on initiatives in farmer management. This chapter examines the way in which process documentation research (PDR) has provided a purposeful instrument to achieve this.

In India, PIM currently shapes large-scale investment in land and water management programmes. This policy shift is rooted in state concerns about the financial viability of irrigation systems. For the past few decades the focus has been on increasing farmers' contributions through enhanced water charges. Thus in 1972, the Indian Irrigation Commission recommended that the water rates should be so fixed that irrigation works do not become a burden on the state exchequer. In 1987, the National Water Policy stated that the water rates charged should be adequate to cover the annual operation and maintenance cost and a part of capital cost of a project. However, by the early 1980s, in Gujarat state the average annual revenue from the water rates covered only about 8 per cent of the annual operation and maintenance costs. In 1985, an expert group of the state's Irrigation Department recommended that water rates should be gradually increased so that by 1991–2 the revised rates would be able to cover 33 per cent of the annual operation and maintenance costs. The recommendation was endorsed by the Gujarat Agriculture Commission in 1988. Lastly, and most importantly, the state government is currently contemplating the introduction of participatory farmer-management in the high-profile Sardar Sarovar Project on the river Narmada which has an irrigated area of 1.8 million hectares.

Despite its consistent policy intention, the fact is that the state government of Gujarat has not been able to revise water rates. This failure of the state has given focus to a number of voluntary organisations and cooperatives which, by contrast, have been able to charge significantly higher rates for irrigation water and recover them from the farmers. In most such experiments, the farmers have actively participated in the management of the system after being convinced of the certainty of adequate, assured and timely availability of water. Indeed, there is now a prevailing consensus that the key to making irrigation water management systems viable lies in promoting farmer management. Thus in June 1995, the Gujarat government passed a resolution 'to introduce [the] Participatory Irrigation Management principle, based on partnership between farmers' associations and Government as [a] "Turnover Programme" for [the] administration and economical management of Government water resources' (DSC 1996). Thirteen pilot projects were identified in different regions of the state in which the management of the projects was to be 'turned over' to farmers after necessary physical improvements had been made. In the case of five pilot projects, voluntary organisations have been involved in the planning and implementation. This initiative presents a major policy innovation.

A parallel initiative came from the department of Rural Development which in 1994 under a national anti-poverty programme financed two rain water harvesting projects in Bhavnagar district in Gujarat. Like PIM this initiative also involves both the development of physical infrastructure and community management through village-level institutions.

The most important shift in policy in both these projects is the shift from a top-down 'blueprint' approach to a participatory 'process approach'. As Korten puts it, 'under a community-based system of resource management, the task of a responsible government agency is not to control all development resources. Rather it is to enable communities to mobilise, control and manage resources more effectively for their own benefit' (Korten 1989). The relative failure of government programmes and the apparent success of NGO projects suggested a need to understand the critical elements in participatory resources management. Furthermore, implementing agencies themselves face a need to explore new ways of gaining knowledge about the context, strategies and impact of their own interventions.

Facing a similar policy need for local innovation in the late 1970s, the Philippine National Irrigation Administration (NIA) developed a tool for the recording and analysis of the implementation of experimental approaches to irrigation management which it called process documentation research (PDR). There too the context was an initiative for community management and the strengthening of a communal irrigation system through new water users' associations.

PDR, an adaptation of participant observation research, provided a means to gain access to the experiences of selected pilot irrigation projects. It generated detailed information on the process of village-level implementation for NIA staff (de los Reyes 1989), providing a 'window' through which to view the 'how' and 'why' of uncensored field experience as the programme proceeds (Shah 1993).

Following the Philippines experiment, the scope of PDR has further widened in that it now aims to provide an increasingly interactive means for agencies to learn about the process aspects of programme implementation. However, the core method of information collection during the course of project implementation (as against *ex post* documentation) remains the same. As argued in Chapter 1, the observation and analysis of the implementation process helps organisations measure the progress of a development activity and assess its impact on the village society and economy. PDR is equally useful in determining the sustainability, viability and replicability of given programmes and approaches.

It is with these objectives that PDR support was provided to the agencies implementing participatory irrigation management projects and water harvesting projects in Gujarat. This support was provided by the Gujarat Institute of Development Research (GIDR), Ahmedabad and financed by the Ford Foundation. GIDR already had, at the time, some experience of *ex post* documentation of processes in a few participatory development projects. In 1993, for example, the Institute worked on two participatory irrigation projects for the Aga Khan Rural Support Programme (AKRSP) in Gujarat. On the strength of this, the government of Gujarat agreed to appoint GIDR as the PDR agency for five of the thirteen pilot participatory irrigation projects in the state. The remainder of this chapter analyses our preliminary experience in conducting PDR in the AKRSP programme, and subsequent adaptations in three of the participatory pilot irrigation projects, and in two rainwater harvesting projects.

Process documentation research in the Aga Khan Rural Support Programme (AKRSP) projects

In Gujarat, AKRSP first experimented with participatory irrigation management in three districts, namely, Bharuch, Surendranagar and Junagadh. Here we focus on work undertaken in the latter two districts. GIDR undertook PDR in selected villages both on schemes to rehabilitate existing irriga-tion systems and on new schemes. The AKRSP programme of PIM began in 1986 with the rehabilitation of an existing percolation tank benefiting two contiguous villages – Rupavati and Devalia – in Surendranagar district. GIDR was invited to document the processes in the Rupavati tank project and the irrigation scheme called the Bandiabelli project. The second project for which GIDR was invited to document the processes was a new lift irri-gation scheme (LIS) constructed in Zadka village in Junagadh district.

AKRSP's general approach involved the selection of villages and programmes and then the formation of a 'village development society' (Gram Vikas Mandal, or GVM) involving participating households as members. With the help of AKRSP officials, the members elect/select society office-bearers – a president, a secretary – and a managing committee to take care of day-to-day administration, the maintenance of accounts, and liaison with AKRSP and other organisations such as the banks. AKRSP emphasises the role of the GVM and its members in programme implementation and coordination with government and other agencies. Moreover, the construc-tion work in these schemes is undertaken with locally available labour and supervised by AKRSP staff.

Given this programme orientation, the PDR concentrated on, first, the NGO–villagers interface including the evolution of water users' organisa-tions and farmers' participation in them; second, various aspects and issues of farmers' contributions towards the share capital and the operation and maintenance fund; and, third, farmers' involvement in the planning and execution of repairs and rehabilitation of the structures.

PDR in participatory pilot irrigation projects

In Gujarat, PDR also has an important role in influencing policy change on irrigation at the state level. AKRSP was instrumental in introducing partici-patory irrigation management through lobbying the Gujarat government. This involved the communication of successful NGO approaches to a wider range of stakeholders in irrigation, culminating in a state-level consultation workshop organised in February 1993 by the government of Gujarat to agree recommendations for the involvement of farmers in irrigation projects. The workshop brought together an unprecedented range of players, including representatives from the government of India, the govern-ment of Maharashtra (a neighbouring state), NGOs, farmers' cooperatives, educational institutions, an expert from the Philippines, the Ford Foundation, the European Economic Community and state government officials. The participation of this wide range of 'stakeholders' was institu-tionalised through a government resolution to constitute a High-Level Working Group chaired by the Chief Secretary. In November 1994, at the instance of the World Bank, the Working Group also included social scientists from the two institutions involved in PDR (GIDR and the Institute of Rural Management, Anand). Finally, in January 1996, the Director of Canals for the Sardar Sarovar Narmada Corporation – a body set up to implement this major irrigation programme – was also included as a member in view of the plans to introduce participatory irrigation management in the Narmada project. This Working Group sought broad consensus on what was viewed as a major policy innovation, initially implemented in thirteen pilot irrigation projects chosen to be 'learning laboratories'. At these sites the programme included the formation of water users' associations, and the organisation of farmers into cooperatives. These tasks were to be undertaken by NGOs.

Following the experience of the NIA in the Philippines, five pilot projects were selected for close monitoring through PDR beginning in October 1995. The specific tasks of the process documenters were: first, to evolve a broadly defined framework for identifying the leading questions and issues and a methodology for data collection; second, to prepare concurrent PDR reports on the implementation of the projects and to discuss them periodically with the implementing agencies and the working groups set up for this purpose; and third, to bring out occasional papers analysing the linkages between different components of the participatory approach of the project.

PDR in rainwater harvesting structures

A similar approach was adopted in relation to the water harvesting project implemented by two voluntary organisations in Bhavnagar district of Gujarat: Kundla Taluka Gram Seva Mandal (KTGSM) and Lok Bharati. The project involves planning and building rainwater harvesting structures such as check dams, gully plugs, earthen dams and percolation tanks to conserve rainwater and to augment dwindling ground water levels. As with AKRSP and PIM pilot projects, there is an emphasis on promoting local management systems, village institutions, and capital contributions (up to 20 per cent) from the benefiting farmers. Equally PDR focuses on the NGO–villager interface, farmer group development and farmers' contributions and involvement in structure maintenance.

In each of these programme contexts it was necessary, before undertaking PDR, to explain the methodology and address apprehension among the implementing agencies (for details, see below).

Methodology

Our process documentation methods have both evolved and varied between projects. Initially, in the AKRSP projects in Rupavati and Zadka villages, the approach involved *ex post* documentation of implementation processes. Both quantitative and qualitative data were collected. The former involved a structured questionnaire which proved helpful in building rapport through house-to-house contact. Qualitative methods hinged on the role of the 'field observers' who were placed close to the project villages. Their task was to build a picture of local communities and the context of programme implementation. It was important to explain clearly the purpose of the PDR to villagers and project field staff alike, and to establish a distinction in the eyes of villagers between the process documentors and implementing agency personnel. Once this was achieved, focus group discussions (FGDs) were conducted to trace the history of development interventions in rural development in the villages. The groups mostly comprised the members from the same community or those who had a similar stake in the AKRSP programmes.

A number of meetings were held in the two villages with the office-bearers and ordinary members of the Gram Vikas Mandals, as well as with non-participating landed and landless households. In-depth interviews were also conducted with the staff of AKRSP and some of the members who owned land in the project villages but did not ordinarily live there.

In the case of the pilot participatory irrigation and rainwater conservation projects, the approach adopted for PDR was more directly focused on events as they occurred. Reporting to the Working Group was therefore concurrent. Methods here involved a combination of participant observation, ethnographic research and survey methods. Since PDR involves collection of information on the project and related activities as they unfold in the field, in each project village a field observer (FO) was placed to facilitate the collection of information. The FO is the last link in the chain of researchers providing eyes and ears of the team at the grass-roots level.

As with the AKRSP process documentation, attention was paid to collecting baseline data from rapid surveys and developing rapport with villagers. Despite good rapport, some difficult issues remained; for example, the question of village selection. Given the impossibility of covering all the project villages, selection is inevitable (four of the thirty villages in the case of the rainwater harvesting project). The question was how to select villages; what happens to important processes that take place in villages that are left out; or who should select the villages – the PDR agency or the implementing agency? We shall take up this and other issues in the last section of the chapter.

Major process documentation research results

In this section we will review some cases to show the results which PDR work has produced in each of the programmes where it has been taken up.

The AKRSP Project

Process research in the AKRSP villages generated some important insights. In Rupavati village, for example, process research highlighted differences in perception between AKRSP staff and participating villagers. In particular, while the NGO's irrigation tank development objectives emphasised participation as a means to achieve long-term improvements in the performance of an irrigation system, villagers gave priority to short-term wage-employment benefits.

The PDR also threw light on the dynamics of intra-village conflict which, in Rupavati, eventually led to the formation of separate institutions for the two principal caste groups. The implications of this for the functioning of the local resource management institutions, for village committees, and for handling the risks of free riding on infrastructural and institutional benefits were also traced through process research.

The process research identified the way in which patron–client systems continued to operate within new 'people's institutions'. On the one hand this led to an unequal allocation of resources as powerful players gained privileged access to benefits; on the other hand, the new institutions were able to effect new challenges to existing structures. For example, PDR work showed how AKRSP was able to challenge the caste dominance of Bharwads over lower status Kolis, and through the structure and procedures of the new 'village development society' (the GVM or Grama Vikas Mandal) institutionalise a new symmetry in village-level caste relations. However, this social balance went along with weak management and allowed other asymmetries, such as between the head-enders and the tail-enders in the use of the irrigation system, to persist.

PDR also helped dispel certain illusions about villager cooperation and voluntarism. First, it was found, in Rupavati village, that cooperative action was usually sustained where enforced by traditional authority structures. Even so, cooperation was always threatened by the inequity between head- and tail-enders – the former contributing less than the latter and appropriating more (in the form of water and agricultural production). Second, participatory approaches had not significantly lowered the peoples' dependence on the state for the management and maintenance of the irrigation system. (Arguably, the increased water charges exacted under the NGO programme reduced the farmers' maintenance obligation.) Overall, process documentation underscored and explained the importance of NGO interventions in initiating, supporting and guiding new water user cooperation.

Methodologically, process documentation in AKRSP projects involved a reconstruction of the history of the implementation of various schemes, and then an assessment of the programme. This assessment analysed (a) the impact of programme implementation on participant households and the village economy, and (b) the logical and causal relationships between processes and outcomes. In view of the *ex post* and evaluative nature of the documentation such an exercise can perhaps best be described as 'process evaluation research'.

Pilot irrigation projects

From a number of points of view, the PDR approach with the pilot participatory irrigation management projects was clearly different from that of AKRSP. First, PDR here was (and is) continuous with the project implementation, being undertaken (at the time of writing) in three of the five pilot irrigation projects. In all three sites (Thalota, Lakshmipura and Chandrawadi) field observers send regular reports on the processes as they unfold in the field. Second, the focus and relevance of PDR information generated was sharpened. It was decided at the outset to generate data on aspects of project implementation as well as on the impact on a cross-section of the village community. Third, process documentors did not attempt to present themselves as 'independent' researcher-observers, but interacted more closely with the project implementation agency. Such interactions also served as a feedback mechanism allowing 'mid-course corrections' to the project as well as to the PDR design. For example, the quarterly PDR reporting format was supplemented by monthly PDR newssheets to provide usable feedback to local agency workers.

Our experience so far shows that implementing agencies, both the government and the NGOs, have given serious attention to issues brought out by the PDR effort. However, it has rarely been clear precisely to whom PDR information is, or should be, directed and tailored. Different actors need different things. While agency field workers value more local and regular feedback on their action, the 'High-Level Working Group' which meets only annually is unable to find time to consider detailed PDR reports. We return to this issue in the last section of this chapter.

Rainwater harvesting projects

The PDR in rainwater harvesting projects which began in late 1995 involved an eclectic methodology. Since PDR began a year after the start of the project, social researchers had to begin with methods such as focus group discussions (FGDs) to capture (retrospectively) aspects of the initial implementation process and the villagers' perceptions of the project. Locally based 'field observers' then collected detailed information on the magnitude of work done by the farmer groups and the modus operandi of both of the NGO agencies involved (i.e. KTGSM and Lok Bharati). In addition to participant observation, the field observers conducted semi-structured interviews with farmers and project workers to help construct the total project story. The aim was to collect data on the tasks undertaken and the procedures followed by the project staff, participant farmers and others. A quarterly report was presented to the overseeing Working Group. To allow feedback, the first draft of the report is prepared in the field in Gujarati and presented to project personnel.

Initially, PDR took the whole project area as its unit of analysis. This made sense because of the logic of planning water harvesting structures on natural water flows. However, given time and resource constraints, it was clearly not possible for one FO to observe and document the activities in the thirty-five or so villages covered. Subsequently, therefore, a few (three) villages were selected in both the agency areas.

The focus of PDR also changed as programme implementation proceeded. Field observers shifted attention from the nature of the schemes implemented (i.e. water harvesting structures) towards the tasks, the procedures and the participation of the farmer user groups involved. Since, in this case, the new water harvesting structures did not anyway generate complex problems of maintenance or system management, PDR had more to contribute in understanding the manner in which farmer groups were able to handle a diversified range of activities in the areas of savings and the management of inputs and improved agronomic practices.

Some issues

To end this chapter we wish to highlight some issues arising from our recent and on-going experience of PDR in Gujarat. In the first place, there has been a contrast between, on the one hand, the *ex post* and evaluative type of process documentation undertaken in a clearly defined area for a discrete NGO client (AKRSP) and, on the other hand, the open-ended, continuous PDR of the participatory irrigation management (PIM) programme addressed to multiple stakeholders interested in more diffuse problems of policy innovation.

A second issue concerns the difficulties when PDR attempts to embrace multiple stakeholders working at different 'levels' – both the 'high-level working group' and the staff of the implementing agency. However, this is necessary if PDR is both to allow effective 'course corrections' and to influence broader policy debate. A singular focus on the working group would mean that PDR outputs and results were greatly underused. In Gujarat, this has been resolved by deciding to present the PDR reports to a standing committee meeting quarterly. The issue of use and user in PDR thus assumes a major significance.

A third issue is whether PDR reports should be explicitly evaluative rather than purely 'documentary'. The importance of critical observation of events during the course of PDR is well recognised in the sense that it can simultaneously facilitate further action in the context of a 'learning process approach'. However, experience shows that the evaluative content in the

PDR is not welcomed by all the agencies, and it risks changing the 'external' status of the researcher.

A fourth issue concerns what unit of observation should be defined for PDR. This was referred to in connection with our experience of PDR in the rainwater harvesting project which covered thirty-five or more villages. The process researchers faced two problems: one was the practical difficulty of covering the extensive watershed area; the other was the inevitably unrepresentative nature of a small number of PDR case studies. Added to this was the agency's concern about fair and representative portrayal of their interventions. PDR on irrigation groups is different in that its unit of study is defined by the jurisdiction of the water users' associations. Nonetheless difficulties arise where the irrigated area, village membership and land ownership do not neatly overlap. PDR is, of course, able to examine the events which occur when project or administratively defined boundaries intersect with the social processes of natural resources management.

Overall these PDR efforts have not aimed to provide generalisable blueprint solutions for participatory natural resources development. Rather they help to evolve organisational capabilities, to learn from one's own experience, and to adapt to specific field situations. In this sense, PDR is an integral part of project implementation on account of its ability to provide constructive feedback to the implementors. Admittedly, the documentation as well as its analysis will be fruitful only when there is clarity about the unit of analysis and the users of the PDR.

As far as coverage and content of the PDR reports are concerned, the methods emphasise an open-ended, inductive research not tied to a priori hypotheses, although this does not preclude use of a broadly defined framework for identifying the leading issues and the methodology for data collection, preferably in close coordination with the implementing agencies.

Acknowledgements

An earlier version of this chapter was presented in a workshop on Process Documentation Research held at BAIF, Pune, December 9, 1996. The authors are extremely grateful to David Mosse for his incisive comments on successive drafts of this chapter and for his editorial help.

References

Appadurai, A. (1989) *Conversations Between Economists and Anthropologists: Methodological Issues in Measuring Economic Change in Rural India*, New Delhi: Oxford University Press.

Basant, R., Kumar, B.L. and Parthasarathy, R. (1994) *Non-Agricultural Employment in Rural Gujarat: Patterns and Trends*, report submitted to Industrial Development Bank of India, February.

Chambers, R. (1988) *Managing Canal Irrigation: Practical Analysis from South Asia*, New Delhi: Oxford and IBH Publishing.

de los Reyes, R.P. (1989) 'Development of process documentation research', in C.C. Veneracion (ed.), *A Decade of Process Documentation Research: Reflections and Synthesis*, Quezon City: Institute of Philippine Culture, Ateneo de Manila University, pp. 21–44.

Dhawan, B.D. (1993) *Trends and New Tendencies in Indian Irrigated Agriculture*, New Delhi: Commonwealth Publishers.

DSC (1996) *Participatory Irrigation Management: A Compendium of Gujarat Government's Orders*, Ahmedabad: Development Support Centre.

Iyengar, Sudarshan (1995) 'An Account of Voluntary Organizations on the Eve of Twenty first Century', in *Gujarati Paryaya*, Volume 1, No. 2.

Korten, D.C. (1989) 'Social Science in the Service of Social Transformation', in C.C. Veneracion (ed.), *A Decade of Process Documentation Research: Reflections and Synthesis* (Chapter 2), Quezon City: Institute of Philippine Culture, Ateneo de Manila University.

Mosse, D. (1994) 'Authority, Gender and Knowledge: Theoretical Reflections on the Practice of Participatory Rural Appraisal', *Development and Change*, 25, 3: 497–526.

Parthasarathy, R. (1994a) 'Policies, Peoples' Participation and the Importance of Documenting the Process of Implementing Irrigation Projects: An Overview of the Experiences of India and the Philippines', paper presented at the Workshop on Process Documentation Research held at the Gujarat Institute of Development Research, Ahmedabad, 11 August 1994.

—— (1994b) *The Process of Implementing Rural Development Schemes and Managing Asymmetries: AKRSP in Rupavati Village*, report submitted to Aga Khan Rural Support Programme (India), Ahmedabad, June 1994.

—— (1994c) *Lift Irrigation Scheme in Zadka Village: Documentation of the Process of Implementation by AKRSP and Some Issues*, report submitted to Aga Khan Rural Support Programme, Ahmedabad.

PRIA (Society for Participatory Research in Asia) (1993) 'Process Documentation in Social Development Programmes', mimeo.

Sengupta, N. (1991) *Managing Common Property: Irrigation in India and the Philippines*, New Delhi: Sage Publications.

Shah, Anil C. (1993) 'Piloting Participatory Development the Philippines Way', in *Wasteland News*, May–July 1993, pp. 19–23.

Sharan, G. and Kayastha, S. (1990) *Lift Irrigation in Panchmahal, Centre for Management in Agriculture*, Monograph series No. 144, Indian Institute of Management, Ahmedabad.

Singhi P.M. *et al.* (1991) 'Rainfall Pattern in Gujarat (1971–1991)', *Working Paper* No. 8, Ahmedabad: Indian Institute of Management.

Veneracion, C.C. (ed.) (1989) *A Decade of Process Documentation Research: Reflections and Synthesis*, Quezon City: Institute of Philippine Culture, Ateneo de Manila University.

4

AN EVOLUTIONARY APPROACH TO ORGANISATIONAL LEARNING

An experiment by an NGO in Bangladesh

Rick Davies[1]

Introduction

This chapter describes what is believed to be an innovative approach to project monitoring, developed in cooperation with the Christian Commission for Development in Bangladesh (CCDB) in 1994. A participatory monitoring system was developed in the course of developing an evolutionary perspective on learning within organisations. The design involved the deliberate abandonment of the use of 'indicators', a central concept in orthodox approaches to monitoring. Instead, the focus of the system is on the identification of significant change as perceived and interpreted by the various participants. It relies on the use of qualitative, not quantitative, information. The approach is inductive, extracting meaning out of events that have already taken place, not deductive, making assumptions about future events. The focus of the system is flexible and adaptive, not fixed. Although the epistemology embedded in the monitoring system is more post-modernist than positivist, the system has proved to be of value to CCDB and has been continued and expanded in its scale of operation since 1994.

The first section of this chapter outlines the methodology by detailing the steps involved in its operation. This is followed by a summary of the state of the monitoring system as of March 1995, a year after the first steps were taken to establish it. A series of contrasts are then made between this participatory monitoring system (PMS) and what are described as orthodox approaches to project monitoring. Questions are then raised about how to evaluate monitoring systems. Finally, an interim conclusion about the value of the experiment is stated, and two issues for further exploration are identified.

The context

The Christian Commission for Development in Bangladesh (CCDB) is a medium sized Bangladeshi non-government organisation (NGO) with almost 550 staff. Its annual budget of approximately US$4 million is funded by a consortium of Protestant donor agencies, in addition to its own internally generated income. Its main programme is the People's Participatory Rural Development Programme (PPRDP), which involves more than 46,000 people in 785 villages in ten districts. Approximately 80 per cent of the direct beneficiaries are women. Development assistance is made available to participants in three forms: group-based savings and credit facilities used to meet the needs of individual households, grant assistance given to the same groups on a pro-rata basis and intended for community-level developments, and skills training, mainly for livelihood purposes. The large-scale and open-ended nature of these activities poses a major problem for the design of any system intended to monitor process and outcome.

In 1994 an experiment in participatory monitoring was conducted with CCDB's PPRDP programme. Implementation took place in four PPRDP project areas in Rajshahi zone of western Bangladesh, where 140 CCDB staff are working with approximately 16,500 people grouped into 503 *shomiti* (associations).

The experiment was the outcome of a voluntary collaboration between the author as an independent researcher and CCDB. CCDB adopted the approach because, according to the Director, it appeared to fit their needs. The experiment was not funded by or encouraged by CCDB's donors, nor was the author in any way answerable to those donors. The author worked with a male staff member from the Training Unit, located in the Dhaka office of CCDB, and saw himself as answerable to the CCDB Director. This immediate allegiance was moderated by a longer-term need to generate information of value to a PhD thesis.

An outline of the process, as implemented

Nine steps were involved. These are outlined in detail below.

Step 1: the selection of domains of change to be monitored

Through the Director of CCDB the author facilitated a brief process whereby Dhaka-based senior staff identified three broad areas or types of changes they thought CCDB needed to monitor at the project level. The number was limited to three in order to keep procedures simple at the earliest stages of what was initially an experiment.

The three types of change selected were phrased as follows:

'Changes in people's lives.'
'Changes in people's participation.'
'Changes in the sustainability of people's institutions and their activities.'

None of these types of change were precisely defined. Their boundaries were deliberately left 'fuzzy'. Initially it would be up to the field-level staff to interpret what they felt was a change belonging to any one of these categories. One additional type of change was included. This was *any other type of change* as judged important by the project-level staff. The intention was to leave one completely open window through which field-level staff could define what was important and report accordingly. In the case of the first three domains of change it was the Dhaka head office staff who had proposed the boundaries or window within which events would be reported.

Step 2: the reporting period

Since the first trial of the method in March 1994 changes have been reported for each of the months from April onwards in each of the four Rajshahi project areas. An experiment was made with fortnightly reporting in April but this was found to be too demanding of staff time, particularly at the head office level.

Step 3: the participants

There were four groups of participants in the monitoring system: (a) the *shomiti* members in the Rajshahi area, (b) the project staff in the Rajshahi area, (c) the senior staff in the head office of Dhaka, and (d) CCDB's donors, particularly those participating in the annual round table meeting (RTM). The structure of their participation determined how the information from the monitoring system was analysed. This is discussed in detail under Step 5 below.

Step 4: phrasing the question

The basis to the monitoring system was a simple question in the following form:

> During the last month, in your opinion, what do you think was the most significant change that took place *in the lives of people participating in the PPRDP project*?

The respondent was then asked to give an answer (written in Bangla) in two parts. The first part was *descriptive*: what happened, who was involved, where did it happen, when did it happen? The intention was that there should be enough information written down so that an independent person could visit the area, find the people involved and verify that the event took place as described.

The second part of the answer was expected to be *explanatory*. The respondents must explain why they thought the change was the most significant out of all the changes that took place in that month. In particular, what difference did it make already, or would it make in the future?

Significance was not expected in any absolute sense, but rather in a relative sense, evident when the various changes that were observed to have taken place in the same reporting period were compared to each other.

It was not expected that the explanation of significance would be objective. On the contrary, it would be a subjective expression of the respondents' values and concerns. The purpose of the explanation was to help bring these values into the public realm where they could be examined, compared and selected.

The process of sampling changes that was involved was *purposive rather than random*. The monitoring system was not reporting the average state of the PPRDP, but rather what was taking place on the outer perimeter of the programme's experience, *the most significant* events. If the reported change was a negative one, then it was a type of change the PPRDP would want to move away from, to avoid in the future. If it was a positive one, then it was a type of change that the PPRDP would want to see become more central to its programme, more typical of its activities as a whole, in the future.

Step 5: the structure of participation

In March 1994 a workshop was held with the senior staff of the four Rajshahi project offices to plan the implementation of the monitoring system. Each project office was told that at the end of each month thereafter they would be expected to report to the head office one significant change in each of the four domains of change. Each project office was then asked to draw up a plan for how their field staff would, each month, identify a range of potentially important changes and how these would then be analysed in order to identify the most important. This change would then be sent by the project office to the head office in Dhaka. For research reasons no requirements or constraints were imposed on who could or could not be involved in the identification of significant changes within each of the project offices. However, it would not have undermined the basic methodology if the head office had imposed a specified common process. For example, all project offices must involve *shomiti* representatives in this process. In this experiment project offices were not told that they had to include beneficiaries, or that they could not include beneficiaries in this process. They were also told they were free to copy from each others' plans if they wished.

Some options concerning methods of selecting from an array of significant changes were outlined, specifically the possibility of using hierarchy (immediate bosses) or teams (of peers) to make the selection of the most significant change out of all they had identified. There was no requirement that the same approach be used in each project area. Nor was there any requirement that the plan individual project offices made would have to be rigidly adhered to thereafter. However, it was insisted that if the plan was changed then the new plan should be made clear to the head office. The central requirement was that however the changes were identified and then selected to be sent to Dhaka, it should be transparent and accountable to those reading the selected accounts. In practice, an average of 15 changes were documented by the field staff at each project office, each month, out of which four were then selected by more senior staff in the same project office as the most significant and sent on to Dhaka.

The same process was repeated at the Dhaka head office. The four sets of four changes (one set of four from each project office) were brought to the head office each month. The task of the head office staff was to select the four changes from the sixteen which they thought were the most significant of all. In other words, the single most significant change in people's lives, in people's participation, in sustainability, and change of any other type. The choice of participants was left to the Director. In practice between four and eight senior staff attended each of the monthly meetings. The process whereby the choice was made by the Dhaka participants was left up to that group. In practice they decided that each participant would rate each story out of ten, and the ratings would then be aggregated to produce the group response. The rating process was preceded by an active group discussion of each account of change. The single requirement was that the group must document and explain their choice, including who was involved in that process, and communicate it back to the staff in the four project offices. In practice, the Dhaka office meeting and discussion took about three hours of staff time per month.

The third level in this process of analysis involved the donors who attended the round table meeting (RTM) in Dhaka in November 1994. By the end of September the CCDB head office had selected twenty-four accounts of significant changes (four domains multiplied by six months). Those changes were collated in the form of four chapters in a report. The introduction outlined the methodology behind their collection (as here), and each chapter thereafter focused on one domain of change (with accounts of change ordered chronologically within the chapters). The appendices detailed an analysis of *shomiti* and staff participation in the monitoring system. It was proposed that donors should read each chapter and select the one change in each chapter which they considered the most significant according to their own values and concerns. As with other participants, they should document the reasons for their choices. In practice, the presence of a wide range of people at the RTM enabled the six months' changes in the first domain, that of the lives of the people, to be analysed in this way by five sub-groups (two donors, one senior staff, one junior staff and one *shomiti* representatives' group).

The structure of participation described above meant that a very wide range of people's life experiences at the *shomiti* level were subject to an iterated process of analysis (choice–explanation–choice . . .) that eventually selected a small number of stories of high value. At each level of the CCDB a range of stories of change were available and subject to a range of interpretations. From amongst these some were selected, retained and forwarded on to the next level in the organisational hierarchy. This process was reiterated from the level of field workers, senior project office staff and senior Dhaka office staff. The process of iterated variation–selection–retention, taking place here within the environment of a single organisation, is the essence of what has been described by Campbell (1969) and others as the evolutionary algorithm.

Step 6: feedback

After each month's changes were evaluated by CCDB head office staff their judgement of the most significant changes, and the reasons behind those judgements, were fed back to the project offices concerned. Similarly, the results of the sub-group discussions at the RTM were also fed back via a plenary session.

The purpose of regular feedback was to allow those identifying the changes in the first instance to take into account the views of CCDB senior staff when in the process of evaluating changes. They could either passively adapt their search for significant change according to the perceived concerns of the head office, or more actively seek better examples and provide better explanations for the significance of the types of changes that *they* thought were most significant. It was intended that if feedback was provided as planned the monitoring system should take the form of a slow but extensive dialogue up and down the CCDB hierarchy each month. In more evolutionary terms it can be seen as a process of co-evolution of interpretative frameworks within an organisational ecology.

Step 7: verification

Those changes that were identified as the most significant of all were precisely those stories where the most effort needed to be invested in verifying the factual details of the event. Verification visits to the sites of the described events can perform a policing function, ensuring that field staff are kept honest in their report writing. They also provide an opportunity to gather more detailed information about the event which was seen as specially significant, and if some time after the event, a chance to see what has happened since the event was first documented (another aspect of impact). Initially follow-up visits were made by the author, with his Training Unit counterpart. Later, in his absence CCDB sent head office staff from the CCDB Information Unit.

The next two steps are optional extras, not ones central to the process.

Step 8: quantification

This can take place at two stages. First when an account of change was documented it was quite possible to included quantitative information as well as qualitative information. Second, it was possible subsequently to quantify the extent to which changes identified as the most significant in one location or zone had taken place in other locations, within a specific period. In the case of one significant change identified in March 1994 (concerning a woman *shomiti* member's purchase of land in her own name) a follow-up letter was sent by the programme coordinator to all ten PPRDP project offices seeking information on the numbers of identical incidents that they were aware of having taken place in their project area in the past year. However, no need was seen to repeat this particular question every month thereafter, as in traditional monitoring systems.

Step 9: monitoring the monitoring system

This step was not essential, but is described here nevertheless. Using records generated by the above process it is possible to monitor changes over time in the proportion of *shomities* and households that the monitoring system is effectively sampling. An analysis in November 1994 showed that after six months of operation accounts of change had been written concerning 43 per cent of the *shomities*. The proportion has continued to grow since then. Similarly it is possible to monitor the degree to which different types of staff (gender, age, position, education) are actively involved in the process of monitoring change, and of those actively involved, with what degree of success. Success in this case is defined as having an account of change being selected as most significant at the project office and Dhaka level. CCDB has yet to do this type of analysis formally, but informally there is awareness of differences between staff in their participation and success. Finally, with an existing data base detailing features of the *shomities* that exist it is possible for specialist staff, such as the CCDB Research Unit, to identify the correlation between changes reported as taking place and objective features of the *shomities*, such as gender, size, and savings levels.

Outcomes so far

CCDB has experimented on two previous occasions with monitoring methods. One system was developed in 1989 by an external consultant and was wholly quantitative in its focus, potentially generating large volumes of tabulated data. It was never implemented. The second was jointly developed in 1993 by a new expatriate staff member working with the staff of the CCDB Information Unit. It focused on the use of rating scales by field staff which were intended to identify the variations in five predetermined types of participation taking place in the people's organisations. It was not implemented.

Although initially planned to operate for the six months until the RTM meeting, the monitoring system described above was continued afterwards, on the instructions of the Director. Delays in the reporting of changes from the project level have not increased, but Dhaka staff were one month behind in their analysis of changes at the time of the last visit to CCDB in March 1995. In January 1995 CCDB decided on its own initiative to extend the system to include three more of the ten PPRDP areas. A training workshop was organised by the writer's CCDB Training Unit counterpart in mid-January, making use of Rajshahi zone staff who have experience with the method. Mention has been made of extending it to other CCDB projects and to specialist support units with the CCDB Dhaka office.

Rather than the contents becoming increasingly identical over time, and the system reaching a form of steady state, new changes have continued to be reported, the most notable of these concerned *shomiti* involvement in an intra-family conflict over contraception use, reported in December 1995. Rather than the project office with the most sycophantic project officer being the most successful (as defined above), the reverse has been the case; success seems to be more correlated with independence of opinion. While the Director had previously identified one project office as the most successful, on the basis of its good credit repayment record, the same project office was the least successful in terms of its ability to generate, through the PMS, a large number of highly rated accounts of significant change.

In contrast to the very limited use made of the output of the CCDB Research Unit over the same period, staff in the Information Unit and the Materials Development Unit of the Dhaka office have made extensive use of stories of change in CCDB publications, videos and educational materials. In addition, project office staff took visiting donor representatives to *shomities* featuring the reported significant changes immediately prior to the RTM in November.

The specific settings of the parameters of the monitoring system have not remained static. In March 1995 CCDB staff were in the process of considering a focus on 'changes in the project management'. In response to demand from the field and head office, the Dhaka staff who give the highest or lowest ratings for a change, which in aggregate has been selected as most significant, had been asked to explain their ratings. As a result of an informal participatory evaluation of the system carried out by the author in February 1995 CCDB staff have become aware of the fact that collectively they see a much wider range of objectives for the monitoring system than were initially conceived by the author and the Director (who wanted evidence of impact). In summary, the monitoring system has survived and is itself undergoing evolution, both in the specifics of its procedures and in its perceived purpose.

The fate of the system is by no means secure, however. Differences of opinion (some explicit) exist among senior CCDB staff, as to the value of the process. In the absence of clear signs of external demand for the information that it is producing, for example, from CCDB's donors, its future is uncertain.

The survival, use and extension of the system exceeded the researcher's initial expectations. Two of the researcher's own expectations were not met. In each account of change that went up the organisational hierarchy there were associated arguments for that choice, added whenever that change was selected. It was expected that middle and especially senior staff would focus on the associated arguments and, through their own choices, selectively reinforce specific types of arguments. Decision premises would thus be subject to evolution over time. In practice it seemed that all levels of staff focused primarily on the descriptions, the empirical events, and gave only passing attention to the criteria of choice used by those below them. Two explanations are possible. One is of 'inappropriate' management style, that middle and senior staff were guilty of micro-management. The other is that the behaviour reflects the staff members' understanding of how the system was supposed to work, that they were in fact required to focus on the events. Explanation of the option of focusing on the criteria of choice used by more junior staff may have been necessary.

The other expectation was that because of the diversity of possible significant changes each set of staff saw they would be required to be critical. They would have to make choices and this would involve comparisons, and thus judgements of relative merit. Participation in meetings at both project office and Dhaka office levels showed that staff could become very animated and assertive in their opinions of relative merit. However, documentation of the choices made, especially the reasons given, only rarely reflected any sense of doubt or criticism, or made any explicit comparison with other changes that had been examined.

A diversity of perceptions offers an organisation choice of direction. One aspect of that diversity is the extent to which the changes that are reported are explicitly negative or positive. Perhaps 90–5 per cent of the changes that were reported were seen by participants as being about positive changes. During the informal evaluation of the PMS it was clear that staff at all levels were quite aware of this aspect of its functioning. Field-level staff explained the risks to their job security involved in more critical reporting of events. Senior staff, less explicitly, indicated concern about donor and governing board responses to negative changes. For CCDB staff the balance of reporting as it existed was functional. It was somewhat disappointing for the researcher who, coming from an academic perspective, initially saw the presence of critical content as a potential indicator of the system's success. Two means of managing balance in reporting were potentially available. One was where negative changes were reported, senior staff could select them and quite consciously provide feedback which lauded the act of reporting as a reason for its selection, along with the specific contents of the report. One problem with this approach, already encountered, is that subtlety of communication style (intended to lower risk) will mean that what is intended as a report of a negative change is in fact not recognised as such by the senior staff. The other means was for CCDB staff to establish a fourth separate and specific domain where 'negative changes' had to be reported. Here the researcher's interests as a researcher meant that this option was not spelled out as clearly, or encouraged as much, as it might have been.

Orthodox versus evolutionary approaches to monitoring

Current approaches to monitoring found in the larger NGOs as well as in bilateral and multilateral aid organisations are heavily influenced by a planning ethos that places substantial emphasis on rationality, prediction and control (Davies 1994). The approach documented in this chapter is in many respects the opposite. Seven differences can be noted.

1   Perhaps the central feature of planning base methods of monitoring is the use of 'indicators', and the need for a *common understanding* about them within the organisation if the monitoring system is to work as intended. That understanding ideally includes:

    (a) the meaning of the 'indicator', what it is meant to represent,
    (b) the worth of monitoring the event represented by the indicator rather than any other possibilities, and
    (c) the existence of the event, did it take place?

    Confusion, especially over their existence or meaning, is seen as a threat to the system's functioning.

    Within the conventional approach it is believed that differences in the subjective perspectives of events and the underlying value concerns of different observers, need to be controlled or ignored.

    Under the evolutionary approach agreement on the meaning of events is an outcome at the end of a process (a month's cycle or more), never final in its form, and subject to revision in the light of new experience. The identification of differences in interpretation is central to the whole process, they are to be brought to the surface and explored, not to be ruled out. In some respects, especially the unfinished and tentative nature of knowledge, the epistemology of the evolutionary approach is a more post-modern outlook, whereas the conventional approach is closer to positivism.

The PMS acknowledges the fact that different sets of values and interpretative frameworks co-exist at different levels within an NGO, and between the NGO and its donors. The NGO field-level staff are in effect creating menus of possible views of the world within broad categories defined by their bosses (Step 1). Their immediate superiors choose from this menu of views and in the process create a smaller menu based on their view of the world. This menu in turn is chosen from by those above them, according to their views. Although clearly located in a hierarchical structure of power it gives significant power to those at the base, more so than under conventional systems. It also still enables the donors to address their concerns without counter-productively imposing their own agendas on the NGO. The senior staff of CCDB did not address gender issues in their original specification of the three domains of change. Nevertheless, a large proportion of the accounts of change that were processed up the hierarchy had very clear issues of gender equity embedded in them. In the round table meeting between CCDB and its donors the most important differences between them in their interpretation and valuation of the significant changes focused on the gender issues within the descriptions of those changes. These were far more subtle and complex than an indicator approach to gender issues would have revealed.

2   Planning-based systems are almost universally very quantitative in their content. Quantitative analysis is based on the ability to enumerate things or events. Enumeration requires a basic assumption about the identity of events being enumerated, i.e. each of six apples is in fact an apple. At the very basic level of counting quantification is about the homogenisation of experience. Differences between apples are extraneous and irrelevant. One apple and one orange can only be added by regarding them both simply as pieces of fruit. Within the daily experience of organisation those events which are countable are summarised by a process of inclusion. Broad swathes of experience are summarised by totals and averages.

The need to summarise is understandable given that most formal organisations have a hierarchical shape. The daily experience of clients and staff at an organisation's base has to be converted into a form manageable by the very small number of staff at the apex. In CCDB the Dhaka office staff had to make sense of 16,500 person years of beneficiary experience of CCDB each year, in the Rajshahi area alone. The question is whether there are not better means of doing this than quantification via indicators.

Within the evolutionary approach *experience is summarised by selection rather than by inclusion*, it focuses on the exceptional rather than the commonality, and it seeks to differentiate rather than homogenise. It is about defining the meaningful edges of experience rather than identifying a central tendency. Within this perspective an observer of a basket of five or six different types of fruit might selectively summarise that experience by reporting that the orange is best, because it is the sweetest. Other observers might selectively summarise the same experience from a different perspective, reporting that the apple should be removed because it has begun to rot and will spoil the other fruit if left there. Philosophising together they might agree that the rot is the more important feature because unattended to it will preclude a range of later experiences, including those of sweetness.

3   Under the planning-based approach to monitoring, events of concern are identified before their occurrence, rather than afterwards. In conventional systems 'indicators' are established at the beginning of a project, and data in the form of statistics are gathered repeatedly throughout the life of the project, concerning the frequency of those events. In more recent revisions of this approach, described as 'process' approaches, by the Overseas Development Administration (ODA) and others, the identification of indicators may be delayed until the project is established, and may in the course of the life of the project, be redefined more than once. The process is strongly deductive in orientation: start with a conception of the desired state and work down from there to what might be the empirical indicators of its occurrence.

The opposite is an inductive approach, where indicative events are abstracted out of recent experience, and this process is renewed with each new reporting period of the monitoring system. Instead of being predictable it is open-ended.

4   In most monitoring systems events of concern are defined by people distant from those events which are to be monitored. Typically the identification of indicators is carried out by senior staff in organisations, either in specialist research units or senior executive staff. In some cases it is made by executives in head offices in other countries, such as was the case with ActionAid's attempt to identify a common set of indicators for its global programme in the early 1990s. Distance can also be in the form of differences in tribe, caste, class, language and education as well as in geography, between those identifying the indicators and those whose lives they are expected to relate to. Reformist approaches have consisted of taking the indicator identification process down the hierarchy, in some cases, to the beneficiaries themselves whose views are sought, through the use of PRA methods. ActionAid, ACCORD and other Northern NGOs have taken this route in more recent times. The problem with such an approach is the difficulty the NGO then finds,

while operating within the conventional framework, in summarising the information produced by a diversity of locally identified indicators.

The alternative approach is to give those closest to the experience being monitored (e.g. the field staff) the right to pose to those above them a range of competing interpretations of those events. The role of those in power over them then becomes to respond, on a selective rather than inclusive basis, to the menu of options provided to them anew each month. Diversity becomes a source of opportunity rather than a conundrum.

5 Normally the analysis of events documented by an organisation's monitoring system is carried out on a centralised basis, at senior levels of the organisation. Typically, field-level workers do not analyse the data they collect, rather they simply forward information up their hierarchies for others to analyse. In the language of computer design this is a serial and central processor based approach. From a Marxist point of view it might be seen as a form of alienation.

The alternative, which can be described as parallel and distributed processing, is seen both in recent approaches to computer design, and in theorising about information processing in biological systems and markets. Information is not stored or processed on a centralised basis, but is distributed throughout the organisation, and processed locally. Staff not only collect information about events but they make their own evaluation of that information, according to their own local perspective.

6 Normally when conventional monitoring data are analysed this is done in a form and location that strip the data of context. Centrally located analysts of tables of statistics sent from field offices are usually well removed from the site of field experience and have a very limited and static predesigned context in which to interpret the meaning of the events that are summarised. Typically few text comments accompany statistics sent up from field workers.

The alternative makes use of what Geertz (1973) has called 'thick, description', closely textured accounts of events, placed in their local context, and where the role of the observer and their subjectivity is visible. In the world of ordinary people these often take the form of stories or anecdotes. Within the evolutionary approach to monitoring outlined here these 'stories' are accompanied by their readers' interpretations.

7 Most monitoring systems are largely static structures. Indicators remain essentially the same each reporting period, and the same questions are asked again and again. The focus remains the same. Those involved in the monitoring system are also seen as unchanging, they simply do their duty. The possibility of independent (constructive or subversive) staff adaptations of and to the monitoring system is denied.

With the evolutionary approach the contents of the monitoring system are potentially far more dynamic and adaptive. (The reality of practice will of course vary from organisation to organisation.) Events reported reflect both a changing world and a changing set of perceptions within the organisation about what is important within their world. Where quantitative data are sought on the incidence of an event found to be significant by a number of levels of the organisation this can be done on a one-off basis, there is no intrinsic need to repeat the same inquiry each reporting period thereafter. Though there are relatively static domains of change being monitored the fuzzy nature of their definition, and the process of definition by current best example, means the overall focus of the monitoring system can move over time. Finally, the monitoring system is simply another arena of organisational life where staff are expected to adapt, but one where their adaptations should be more visible, and therefore open to more deliberate and selective reinforcement.

Evaluating monitoring systems

The evaluation of a monitoring system is as problematic as the task of evaluating the impact of CCDB's activities on the lives of its beneficiaries. The month to month functioning of the monitoring system is an event subject to observation and interpretation just as much as that of the woman *shomiti* member who purchased land in her own name. There are multiple observers with varying criteria of concern, some within CCDB and some outside. All are making judgements influenced by their current context and past history. What weight should each of those judgements be given and how can they be aggregated or summated?

An approach based on an evolutionary epistemology would start with the fact that the system has survived and treat that as evidence of its value in an aggregate sense. It has some degree of fitness with its environment. But this is a minimalist judgement. In the words of Belew (1991: 9), 'the dumbest smart thing you can do is stay alive'. Associated with this complex recurrent event called a monitoring system are the many interpretations of its meaning, including the values and preferences which sustain it, some of which are more prevalent than others, and whose prevalence changes over time. Success, i.e. some form of aggregate value, might also be judged, though with much more difficulty, by identifying changes in prevalence of these sustaining values. It could be argued that sustainability on this level is more important than the mere survival of a set of organisational routines known as a 'participatory monitoring system'. But on the other hand it could be argued that those values need some form of substantiation to be fully meaningful and thereby to survive as a set.

While the CCDB monitoring system shows that the structured and public exercise of choice is a means of surfacing and managing some values within

81

an organisation, it remains to be established how such a process should be structured when there is a need to evaluate an intra-organisational event in a wider social and organisational context. Donors, government, consultants, and academics all may have an interest in the CCDB experiment, yet they are much more independent of each other than staff within a single hierarchically structured organisation. One possibility, yet to be explored, is the use of a market as a means of summarising judgements. As with the use of hierarchy, its value could be improved by making information about events taking place in that market more transparent and public.

Conclusions and issues remaining

The monitoring system that has been developed has successfully addressed the needs of CCDB as an organisation to monitor the impact of its activities. The concepts underlying its functioning are also relevant to the wider issue of how best to analyse qualitative (as distinct from quantitative) data, and how to do this on a participatory rather than solitary basis. The approach proposed suggests that a key issue is inter-subjectivity, the extent to which different observers of events or phenomena agree and disagree with each other. In this chapter the focus has been on the perception of change and of differences between those perceptions. The associated practical issues are how to process such information and at the same time how to represent it. It is suggested that iterated processes of variation–selection–retention, i.e. the use of an evolutionary algorithm, are of value in dealing with both tasks.

If this approach to organisational learning is to be developed two areas need more in-depth exploration. One is the concept of public and private domains of knowledge in organisation. Studies of simulated multi-agent systems have suggested that the capacity for learning involves a balance between order and disorder (Kauffmann 1994). The public domain of knowledge in an organisation could be seen as highly ordered, a domain containing 'that which everyone knows that everyone knows' and one where areas of agreement on meaning provide a further level of order. Many more private domains also exist in organisations, and in their plurality and difference can be seen to embody more organisational disorder. Many of these private domains are likely to remain so, because of the contradictions in interests which could not co-exist in public. What is the role of monitoring systems in relationship to these public and private domains? Widened agreement on an issue, and awareness of that agreement, will almost by definition assist successful joint action by people holding those views. Increased awareness of differences is more ambivalent. It may enable innovations in organisations' objectives and practice or it may threaten the very cohesion, and survival, of the organisation. The question is then 'How can these risks be managed creatively?'.

The second issue is that of internal and external demand for information.

The CCDB experiment took place because of internal perceptions of need, but that awareness included knowledge of donors' views and concerns. The impact of external demands for information is mediated not only by internal processes of interpretation within NGOs, but also by processes of interpretation within their donor organisations. Variations clearly exist in the extent to which donors see their demands as legitimate or questionable (e.g. ODA versus Christian Aid), and burdens to be minimised or lightened versus requests which are potentially educative and enabling. Emphasis also varies in the extent to which they see themselves as seeking confirmation of pre-established expectations versus identification and understanding of news about innovation. The fate of developments such as those within CCDB will undoubtedly be influenced by how these tensions are resolved. In order to understand and possibly influence this process attention will also have to be directed by researchers and consultants to the nature of information demands within donor organisations, to the influences on these demands, and how they can be managed creatively.

Note

1 Centre for Development Studies, Swansea SA2 8PP, Wales, UK. Fax/phone: 44 223 841367, E-mail: rick@shimbir.demon.co.uk, http://www.swan.ac.uk/cds/rd1.htm

References

Belew, R.K. (1991) 'Artificial life: a constructive lower bound for artificial intelligence', *IEEE Expert* 6, 1: 8–15.

Campbell, D.T. (1969) 'Variation, selection and retention in socio-cultural evolution', *General Systems* 16: 69–85.

Davies, J. (1994) 'Information, knowledge and power', *IDS Bulletin* 25, 2: 1–12.

Geertz, C. (1973) *The Interpretation of Cultures*, New York: Basic Books.

Kauffman, S.A. (1993) *The Origins of Order: Self-organization and Selection in Evolution*, Oxford: Oxford University Press.

5

IMPACT ASSESSMENT, PROCESS PROJECTS AND OUTPUT-TO-PURPOSE REVIEWS

Work in progress in the Department for International Development (DFID)

Anne Coles and Phil Evans, with Charlotte Heath[1]

The emergence over the past 10 years or so of process projects as an increasingly major part of the DFID portfolio has coincided with, and been symptomatic of, a gradual change in thinking about how the effectiveness of international development assistance can be enhanced. An accompanying trend has been an increasing concern with results and therefore with the assessment of impact. These trends have developed against the background of a growing awareness of the need for development assistance to be people-centred and for projects to be locally owned in partnerships which, whenever possible, include primary stakeholders.[2] In common with many other development agencies, DFID has focused significant attention on reviewing and reformulating its project cycle management and monitoring and evaluation procedures to more closely reflect and support these lines of thinking and so promote greater effectiveness in its work.

This chapter provides a glimpse of 'work in progress' in DFID, with a particular emphasis on recent experience in the conduct of the new output-to-purpose reviews (OPRs). These aim to take stock, at the mid-point in the implementation of a large project or programme, of progress towards the achievement of its developmental purpose. A key feature of this approach is the recognition of the importance of placing impact assessment within the context of the wider development process, as an integral supportive component rather than a parallel stream of information extraction and audit. Two recent case studies are drawn upon to illustrate current trends and developments. Although the emphasis of the chapter is on one specific part of the project cycle (ongoing impact assessment), the issues raised are in many cases of relevance to the conduct of process projects in general.

OPRs are replacing what used to be called mid-term reviews (MTRs). These have been a long-standing feature of ODA/DFID approaches to project management.[3] While they have been a useful tool in assessing progress, they have been less than satisfactory as opportunities for learning and lacked a sufficiently developed structure to strengthen stakeholder ownership and promote broad-based participation. MTRs often veered towards being seen as supervision exercises, rather than structured, shared lesson-learning events. The emergence of a greater emphasis on process projects, reflecting an increasing preoccupation with human development and institutional reform and a declining emphasis on technology-led inter- ventions, prompted reflection and rethinking of approaches within the organisation. OPRs are intended to be integral to, and fit comfortably into, the routine project management cycle. This has been taking place within the context of a general overhaul of project cycle management (PCM) proce- dures, including a streamlining and sharpening up of the use made of the logical framework as a design and management tool.[4] The wider context for this process has been the challenge, widely felt within the development community, to capture and demonstrate aid effectiveness and make better use of increasingly scarce resources.

The wider PCM context for the development of OPRs is of central impor- tance in moving towards a means of capturing progress towards the achievement of developmental objectives in ways which are supportive, rather than disruptive, of the flow of the project cycle: stakeholder involve- ment and shared ownership are critical factors for the success of this approach, along with the demystification of impact assessment work and the importance of allaying suspicions that it remains essentially a policing operation. The challenge in relation to the latter point is to demonstrate a genuine commitment to partnership, without conceding the necessity of accountability to the British public for the way in which DFID is spending public funds.

How do OPRs fit within the DFID's work? Conceptually, the OPR, as its name suggests, is closely linked to the log-frame. Its main function is to assess progress towards achievement of the project's objectives (the purpose and goal) and it re-assesses assumptions and risks. In terms of DFID project cycle management, OPRs differ from other assessments. Annual monitoring typically focuses on progress at the level of activities to outputs, while OPRs concentrate on progress at higher levels of the log-frame. Project completion reports and formal *ex post* evaluations occur at the end of a project's life, while OPRs have the advantages and limitations of taking place during it. In terms of our policy information marker system (PIMS), OPRs also provide an initial assessment of progress against these markers.[5]

OPRs are team events which aim to involve key project stakeholders and, at least for the final part of the review, the main DFID interdisciplinary team involved with the project. They conclude with recommendations for the way forward for the remainder of the life of the project and may contribute material more generally on DFID's aid effectiveness and impact.[6] OPRs are considered to be of particular importance in process projects because of the need for an internal learning mechanism, and to enable outputs to be modified or new ones created during implementation, and assumptions to be tested.

An OPR aims to provide a strategic overview of progress within a DFID funded project or programme towards the delivery of its outputs and the extent to which these will lead to the attainment of its purpose. In most cases, impact measures yield only limited data for a review of this kind, but emerging achievements can be identified and used as a proxy for likely development gains. An OPR creates a space within the project cycle for stock to be taken of the 'big picture', for the aspirations of a project or programme to be reviewed in the light of immediate experience, and for the assumptions and risks informing the original design to be re-assessed. Where changes are required, these can be made in the light of the review: adjustments are often made as a result of OPRs to outputs, indicators, and assumptions. In some cases, changes may also be required at the purpose level, though this would be a signal that something was seriously amiss with the original design.

An OPR is intended to be a timely stocktaking exercise, at a point within the project cycle where enough has been done for emerging trends to be detected, but things are not so advanced that changes can no longer be made. DFID projects are typically three to five years in duration, often with extensions, so that the OPR takes place two to three years after the start. In this respect, OPRs are both an impact assessment and a management tool, facilitating an early assessment of likely developmental outcomes and an opportunity to reassess project design, implementation procedures, and underlying assumptions to increase the prospects for success. By involving all key stakeholders, OPRs also create a forum for differences in view to be explored and a common purpose to be strengthened and reaffirmed, and for a shared lesson-learning process to be promoted, linking primary stakeholders in the field through to DFID senior management, via key recipient country and DFID players in the implementation process.

The OPR concept is relatively new in DFID and a 'state of the art' is still being evolved. A number of key questions remain to be explored, concerning both the costs and benefits of the approach, and the means by which adequate levels of stakeholder participation can be assured.

Assuring adequate levels of stakeholder participation implies a willingness to share control over the process, and steps are being taken to explore ways in which this can be done. Emerging experience from DFID's NGO direct funding initiative (DFI) in Eastern Africa suggests that, under the right circumstances, a willingness to allow others to set the pace in impact assessment work, particularly in the context of process projects, can yield significant dividends. Not only can this serve to strengthen participation and shared ownership, but it also allows space for unexpected findings beyond the confines of the log-frame.

A good example is a joint review undertaken of the Oxfam Pastoralist Development Project in Wajir District, northern Kenya, where the formal review mission was preceded by extensive consultations between the field implementation team and groups of primary stakeholders and the development of terms of reference for the review by the implementing agency, rather than DFID. This approach ensured widely shared ownership and commitment to the review process, and also enabled lessons to be drawn from the review beyond the boundaries set out by the logical framework. An important revelation, for example, was the realisation that the work undertaken by the project in supporting the development of pastoralist associations cutting across clan lines might not only help to strengthen the security of livelihoods but also promote greater social integration and help reduce conflict. The project was bringing together competing clan groups, in a precarious natural environment, across traditional lines of differentiation in pursuit of a wider common purpose, which many members of the community believed would reduce the likelihood of conflict in future periods of crisis.

This example highlights how the introduction of OPRs, reflecting the wider aspiration to get a more solid grip on the difference development assistance makes to people's lives, and the development of a variety of approaches in the field, has already enabled DFID to identify a number of key questions and emerging lessons in taking this work forward.

There is no blueprint for the design of DFID OPRs. Some are more modest in scope than others, and with the exception of a number of DFID funded NGO initiatives, they are not usually required for projects spending less than £250,000. Typically they might involve the following: an independent assessment of the project's achievements to date, the involvement of all project staff in reviewing the log-frame, a field visit by a multidisciplinary team from DFID and their participation in workshops with project staff and key secondary stakeholders to review findings, arrive at conclusions and provide the strategic way ahead. The form that they take depends very much on the size of the project, whether it is seen as having strategic importance, whether it is innovative with the potential for replication, or whether it shows signs of running into problems. Importantly, it depends on the extent to which the project already has an established means of monitoring its progress towards its purpose and a means of capturing its first results. Some south Asian NGOs have established monitoring and evaluation units and undertake considerable operational research. Other projects are much poorer in this respect. The two case studies presented here are very different and highlight variable approaches followed in planning for and conducting OPRs. One is a fundamental review of a project where there was very little in the way of data collection systems. The other describes a project where the main project implementers already had their own systems for assessing impact, and systems are being developed at an early stage to build on this advantage in ways intended to support the wider implementation, lesson-learning, and institutionalisation process.

Case study 1: the Faisalabad Area Upgrading Project (FAUP), Pakistan

The Faisalabad Area Upgrading Project operates in four pilot areas in Pakistan's third largest city. The project has two constituent elements: a 'process' component, and a primary infrastructure one.

The process component of the project aims to develop a sustainable capacity within local communities to participate in the design, financing, implementation, and operation and maintenance of physical and social infrastructure improvements. It also aims to promote income earning opportunities for slum dwellers and to improve the status and meet the socio-economic needs of women. The project involves the capacity strengthening of local government institutions, primarily the Faisalabad Development Authority (FDA) but also parts of the Faisalabad Municipal Corporation (FMC) and the Water and Sanitation Authority (WASA) to plan and manage a participatory approach to the development of low income urban areas.

The primary infrastructure component involves improvement of primary infrastructure for water supply, sewerage, drainage and solid waste disposal in order to increase capacity to allow the extension of services into low income areas.

The project has shown a slower than anticipated process of evolution, taking more than four years from 1993 to achieve a project on the ground with functional capacity. The recent output-to-purpose review (OPR) was therefore a careful but constructive review of what has been seen as a problem project. Agreement to a second phase of the project was contingent on the recommendations. The context, process and main lessons learned are outlined below.

The key issues for the OPR were:

- the extent to which the community organisational development components of the project were bought into or owned by the Government of the Punjab and other key local government institutions;
- progress with the implementation of multi-sectoral (education, health, micro-enterprise, infrastructure) activities;
- the perceived dilemmas posed by the very participatory approach taken at community level and the tensions between this and the hierarchical, highly bureaucratic approach of the government agencies involved;

- issues of institutional complexity and responsiveness to community demand;
- managerial issues, connected with the project management unit, which was experimental, and the roles of the external consultants and DFID;
- the very low morale amongst PMU staff for the last two years at least and the high staff turnover; gender problems amongst staff particularly affecting the female social organisers;
- the very weak monitoring and evaluation by the project, reflected in little systematic impact data at primary stakeholder level and minimal baseline data;
- the difficulties of getting quality consultants (local and UK) to live and work in Faisalabad.

It was agreed that the OPR should broadly review achievement of project outputs and progress towards meeting the project purpose. In view of the known difficulties and gaps in information it was also decided that there should be a number of pre-OPR activities which would feed into the formal review: an external participatory impact assessment (PIA) of primary beneficiaries to be carried out by an external social development consultant and ActionAid (Pakistan), and reviews of institutional and micro-enterprise issues.

The overall approach taken in the PIA was one of participation of both primary and secondary stakeholders. Central to this was the participation of those people whom the project is intended to benefit. A key feature of the approach was that it was reflective and flexible. The PIA consisted of three main components: a participatory survey and analysis of project impacts as perceived by primary stakeholders; an analysis of impact information available from existing secondary data by sector; and the development of a participatory methodology for the impact assessment which may be used in other urban contexts in Pakistan.

ActionAid's planning of the PIA took a couple of weeks. It involved key project staff, management, visits to the four project areas, workshops with project staff and an intensive internal team exercise to develop the study process and methodology.

The following important lessons were learned from the (PIA) planning process:

- 'Ownership' of the project should not be taken for granted but must be built. Ideally the approach to the PIA should be developed with full consultation with the different stakeholders. Stakeholder analysis should be undertaken during the initial planning stage of impact assessment to ensure that those with a stake in the PIA are identified and involved in the early stages.

- Managerial staff need to be involved from the outset in discussion and planning of impact assessments and to define a clear role for themselves in the various stages of the study. Regular meetings to review progress and findings are critical.
- Participatory workshops which are well planned and carefully managed can be an effective tool for developing commitment and ownership amongst project staff.
- PRA techniques can be used to good effect to create an open, relaxed and informative environment.[7]

Field work took place over a four week period with a further week for final data analysis. Further lessons were learned from the implementation and field management of the PIA:

- The adoption of a collaborative, rather than external team, approach, including during the field work stage, has considerable advantages which outweigh concerns about possible bias.
- Transparency about potential bias and agreement on ground rules from the outset mean that potential difficulties and risks are identified and managed.
- Impact assessments can be undertaken by teams with little or no previous experience of work of this kind, if well managed.
- Shared management amongst team members develops shared ownership and responsibility, and enhances the quality of the assessment.
- Reflecting on and documenting experiences and lessons on a daily basis, and reviewing overall experience collectively at the end of the study, maximises learning and individual and team development
- Time must be allowed for feedback to the communities involved, to reach a broader audience and capture more ideas.
- Ensuring that translated material is widely circulated to communities in advance of meetings helps to stimulate debate.
- Findings should be analysed in the field as they emerge.

The formal stage of the OPR followed the PIA. The eight person DFID review team consisted of the desk officer as team leader, with relevant sectoral advisers, and an input from a locally engaged programme officer who had known the project since its design phase. For a project that was designed to promote gender equality the uneven balance of the team (with only two women) was not ideal. The team was facilitated and supported generally throughout by the former head of ActionAid in Pakistan who had participated in the PIA and thus provided valuable continuity.

The OPR covered eleven working days. A draft report was produced and shared with project staff after the fifth working day in Faisalabad. That report was then amended in the light of comments from staff and amended again following meetings with FDA/FMC staff in Faisalabad before presentation to the provincial government in Lahore. Further amendments were made to take into account the government's views.

OPR team members used secondary sources, and drew particularly heavily on the results of the PIA and the institutional review. A number of formal meetings were held with government agencies. One day was spent visiting all four project areas to give grounding to the PIA findings and one day was spent on a workshop to examine the project's performance to date and its strengths and weaknesses, and to identify the way forward. The workshop was attended by federal and provincial government representatives, project staff, staff of FDA (including the Director-General) and FMC, four consultants, the OPR team members and two Community Group representatives.

The following lessons were learned from the OPR:

- Separating the review of progress towards achievement of project goals, and having it carried out by an independent, non DFID team, from the formulation of recommendations for the project's future involving DFID stakeholders, helped strengthen the objectivity of the review and helped allay fears that it would be a policing exercise.
- Local knowledge, and ideally the inclusion of local experts in the team, is indispensable.
- More time to meet members of community groups, as well as community members who were not participating in the new groups, would have been desirable, but time constraints throughout the review were severe. Two community group members did attend the one day workshop but others were excluded because English was the working language.
- Meticulous advance planning is both essential and valuable, and well worth the effort required.
- Vision versus task: the need to begin with the big questions and not be mesmerised by production of the document. The balance between discussion and drafting must be precisely struck.
- The importance of reflection and the identification of next steps: the need for reflection on the process to draw out and disseminate lessons learned, an unpressurised session near the end to agree what happens next and who will do what by when, and a timetable of next steps to be communicated to partners and project staff as soon as possible thereafter.

The Faisalabad review was undertaken by a large team, and was costly in both time and human resources.[8] Given the importance of the project, however, this was considered justifiable. As the listings above suggest, the exercise was also a rich source of important lessons of value beyond the project itself. It was recognised, however, that this level of input may not be justified in every case, and that projects must be assessed and judgements made about the level of resources to be devoted to reviews and a framework developed to undertake this.

Case study 2: the Ha Tinh Poverty programme, Vietnam

Ha Tinh, in north-central Vietnam, is one of the country's poorest provinces, with an estimated 70.9 per cent of the population living below the income poverty line, compared with an average of 57.2 per cent for rural Vietnam as a whole. The Ha Tinh Poverty programme was approved by the ODA (now DFID) Projects and Evaluation Committee (PEC) in March 1997. The programme provides support to the continuation and further development of three NGO poverty reduction projects in four districts in the province. The programme framework seeks to support increases in benefits to poor people on a sustainable basis, and to draw lessons from project experience which can inform more widespread poverty reduction work in the province as a whole.

The constituent projects are being implemented by ActionAid, Oxfam, and Save the Children Fund, and are distributed among the four districts of Can Loc, Cam Xuyen, Ky Anh and Thach Ha. Project components include support to savings and credit schemes, rural infrastructure (including irrigation schemes and construction of sea dikes), food production, environment and natural resource management, and community-based primary education. Firmer links are to be established with the district and provincial administrations to promote lesson learning and encourage the replication of successful approaches.

Project-logical frameworks have been developed for each of the NGO projects, and an umbrella framework established for the programme as a whole. At the programme level, the goal is that poor people in Ha Tinh province benefit from sustainable economic development and lessons learned from the programme being applied elsewhere. The programme purpose is that district and provincial authorities and other development organisations apply lessons learned from innovative poverty projects being run in the participating districts.

A significant feature of the programme is a firm commitment to ongoing impact assessment work, aimed at measuring progress towards sustainable poverty reduction and learning lessons from project implementation for wider replication. Two per cent of the total programme budget has been earmarked for this work and an innovative advisory and support role defined for DFID's Evaluation Department (EvD), from the beginning of the implementation cycle.

PEC proposed that EvD assist in revising the programme framework at the purpose level in particular, to ensure that both the direct and indirect benefits are adequately captured. This work should be undertaken in consultation with the participating NGOs and other key stakeholders. In interpreting this brief, EvD considered it important that a system be devised which would enable both expected and unexpected outcomes and effects to be identified and assessed.

During the course of discussions with the programme partners it was agreed that rather than seeking to identify specific measures for poverty reduction and institutional lesson learning, it would be more appropriate to set out a general framework which would ensure that the various information and lesson learning streams generated by the constituent projects are systematically integrated and key programme measures and findings distilled out. This should be undertaken in a manner supportive of the achievement of the programme purpose (and, eventually, goal), and the longer-term needs of effective summative, as well as formative, programme evaluation.

All three NGO partners have evolved sophisticated approaches to monitoring and evaluation, which taken together provide a sound and adequate basis for approaching the key impact questions which need to be asked, both within the context of the individual projects themselves and for the programme as a whole. The style and emphasis varies from one to another, but all are committed to self-critical and transparent approaches and to the use of quantitative, qualitative, and participatory methods. Where information gaps become apparent, as a result of the joint analyses undertaken by project partners, these will be filled through supplementary work, financed from the central monitoring and evaluation funds set aside for this purpose in the programme budget.

Rather than seeking to fine tune existing approaches, or superimpose a separate system at the programme level, the challenge identified was to create a mechanism to facilitate the regular pooling of findings, insights and experience gained from the field to enable the 'big picture' to be distilled out at provincial level, drawing on the strengths of each of the constituent information streams, as well as helping identify key information gaps of interest to all programme partners which commissioned work may help to fill. This would fit more comfortably into the programme commitment to stakeholder ownership than a prescribed set of impact measures being established at the programme framework level, where ownership and accountability are less easy to ascribe. This does not imply any diminution of the significance accorded to the programme-level framework, particularly at the purpose level, but should instead help pave the way for a genuinely shared commitment to its achievement.

The approach should also have the merit of contributing to the establishment of stronger institutional linkages between the NGOs and the district and provincial layers of government, and help to define the role of the Ha Tinh Province NGO Committee, which has recently been established by the Provincial People's Committee. A key focus of the discussions held between all participants during the mission was the negotiation and defining of the role of the province in programme coordination, reinforcing the importance of the monitoring and evaluation framework in underpinning this and ensuring an active engagement of provincial officials in programme delivery and lesson learning.

As the programme progresses, indicators at the purpose level of the programme framework would be refined, with the annual workshops providing a focal point for this work. These would cover three main areas of concern: the direct impact of project interventions on the livelihoods of poor people (including consideration of increases in productivity and improvements in livelihood security, the distribution of benefits, gender concerns related to implementation and the distribution of benefits, and the management and sustainability of the natural resource base); approaches to implementation and other process factors (including the nature and quality of relationships with participating communities) and the capturing of lessons learned; and measures of institutional lesson learning at the district and provincial levels and the extent to which good practice identified during programme implementation is being replicated more widely.

As well as aggregating lessons learned on an annual basis, the workshops would also present an opportunity to share good practice in impact assessment work, and refine the tools and methods being used by all programme partners.

The central funds set aside for programme monitoring and evaluation, administered by DFID's South East Asia Development Division, would be made available to provide facilitation support (if required) for the workshops, and funding for supplementary studies and output-to-purpose (OPR) and end of programme (EPR) reviews.

The proposed OPR will focus on the programme level, and will draw, among other things, on whatever formative evaluation work has been done by the NGOs and other relevant sources of information, including findings from national welfare and consumption studies. The OPR will also provide an opportunity, if required, for specialist expertise to be drawn in to assist in consolidating and synthesising impact assessment data from all parts of the programme and gaining an overall measure of progress towards the achievement of programme objectives.

The EPR will draw preliminary conclusions about the extent to which programme objectives have been met, and sustainable benefits are likely to be achieved, and will also assist the province, districts, and participating NGOs to identify needs for further poverty reduction work in Ha Tinh. It is also anticipated that lessons learned from Ha Tinh will be of interest outside of the province and will be widely disseminated in Vietnam and elsewhere.

Emerging lessons and challenges for DFID

It is already clear that OPRs, if they are completely to fulfil their potential, demand a more rigorous approach to assessing emerging impact than has been the case with previous approaches to formative review work. The challenge of making up-to-date, reliable, and 'objective' information available in time for the review exercise has raised both methodological and resource questions. How much time and effort should be spent on this work, compared with the 'real' work on project implementation? Are these two activity streams in conflict with one another, or can they be brought together in a mutually supporting way? Are 'outsiders' needed to collect and analyse data to ensure a dispassionate view? What is the role of judgment and experience in drawing lessons? Where stakeholder perceptions differ, how can these be reconciled? To what extent are or should judgments about impact be negotiable?

The issue of stakeholder participation is of critical importance, but, again, important questions remain to be answered. How wide does stakeholder participation need to be? Conventional client surveys will seldom be sufficient. Potential beneficiaries (who are presently non-participants), as well as any unexpected beneficiaries, need to be included. Otherwise opportunities to identify the full range of project outcomes will be missed.

Another clear need, arising from the recognition that there are always many stakeholders in any development process, is that for disaggregated data, capturing the views of and emerging effects on women and men, rural and urban dwellers, rich and poor, old and young, etc. Without the insights which disaggregated data can provide, the scope for fine tuning the project's performance is considerably reduced.

The logical framework, particularly in the revised form in which it is now being used by DFID, is a powerful design and management tool, but care must be taken to ensure that it does not become a straitjacket, or blind implementation teams to realities beyond the boundaries of its framework. Effective impact assessment work, particularly in the context of process projects which are, by definition, evolutionary, must be able to identify unintended outcomes. For this, getting one's head up above the log-frame is essential. The unexpected can include both positive and negative outcomes, spin-off and ripple effects both within and beyond the boundaries of the project, including the effects of the informal dissemination of information and knowledge beyond the project area. Indeed, one of the most powerful signs of project success and an indication of future sustainability is when outsiders spontaneously adopt and replicate some of its elements.

Obtaining information is clearly only a first step. If OPRs and other ongoing means of impact assessment are to be of real benefit, the assessment process must be supportive of efforts to operationalise findings and turn them into effective action on the ground. A learning culture among project management and staff, as well as established feed-back loops for translating lessons from the field into decision-making, are among the pre-requisites.

OPRs are increasingly regarded as a constructive means of lesson learning for project and programme stakeholders. But how effectively are they being used as a way of learning lessons throughout the organisation? DFID is working hard to devise the MIS systems which will enable this to be possible. Successful pilots have been established on a divisional basis. The basic OPR summary reports should increasingly be accessible electronically but in the meantime there is reliance on personal networks.

Many questions remain to be resolved, and more will no doubt arise as experience in the field and in the use of OPRs is gained. There is little question, however, that the pursuit of improved approaches to process-based lesson learning and impact assessment will continue to be central to DFID's work as we move into the twenty-first century and continue to support international efforts to respond to the challenge of global poverty.

Notes

1 The authors are members of DFID's Social Development Advisory Group. Anne Coles is Senior Social Development Adviser responsible for coordinating social development advice to DFID programmes in Asia, Latin America, the Caribbean, and the Pacific; Phil Evans is an adviser in DFID's Evaluation Department; Charlotte Heath is Social Development Adviser for West Asia. Ann Keeling also contributed to the Faisalabad case study. The views expressed in this chapter are those of the authors alone and not those of DFID.

2 DFID defines 'primary stakeholders' as those whom the project is intended to benefit.

3 Following a change of UK government in May 1997, the Overseas Development Administration (ODA) was transformed into a Department of State, to be known as the Department For International Development (DFID).

4 The logical framework attempts to provide a summary, in hierarchical form, of the flow of logic by which the project seeks to achieve its developmental objectives. Activities (and other inputs) are intended to produce a set of outputs, which together should contribute to the achievement of the project's developmental purpose. The purpose, in turn, should contribute (along with other developments) to a higher-level developmental goal. The name output-to-purpose review is derived from this framework, and reflects the level at which such a review addresses issues of project design and implementation. The key question addressed in OPRs is the extent to which a project is likely to deliver its outputs, and the extent to which the assumption that these will, together, lead to the achievement of the purpose is still valid.

5 PIMS provides a simple tracking device for monitoring expenditure allocations against key developmental objectives.

6 For example, the findings of selected OPRs are now being incorporated into evaluation synthesis studies to help make lessons from ex-post evaluation work more relevant to current concerns.

7 It was also learned that the presence on the team of a young child (of a working mother) can have a very positive effect in reducing tensions and providing welcome social diversion.

8 Including the need for a significant level of administrative and secretarial support.

Part 2

PROCESS MONITORING IN INTER-AGENCY CONTEXTS

6

PARTNERSHIP AS PROCESS

Building an institutional ethnography of an
inter-agency aquaculture project in Bangladesh

David J. Lewis[1]

Introduction

This chapter presents research into inter-agency project relationships. A
process view of projects was central to the methodology which was adopted
during the study and this generated a set of distinctive insights and problems. At a conceptual level, ideas about process also contributed to the
widening of our understanding of the nature of 'partnership' between agencies in projects. The roots of the present case study can be traced back to
research undertaken during 1990–2 by the Overseas Development Institute
(ODI), which investigated government and non-governmental organisation
(NGO) linkages during efforts to promote technical change in the agricultural sector using case-studies collected in Asia, Africa and Latin America.
The research concluded that while collaboration between NGOs and
government agencies was certainly taking place and in many cases generating potentially useful new approaches and insights, there was no
straightforward 'functional' division of agency roles and that social, political and historical contextual factors in different countries were crucial
determinants of linkage effectiveness (Farrington and Bebbington 1993).

The South Asia portion of the research, which was documented in
Farrington and Lewis (1993) attracted attention from one of the
Consultative Group on International Agricultural Research (CGIAR)
centres, the International Centre for Living Aquatic Resources Management
(ICLARM). ICLARM decided to develop with ODI a joint research project
in order to study inter-agency linkages in aquaculture. With the assistance of
'holdback' funds from the then Overseas Development Administration
(ODA), a joint project was designed to build on the ODI research using
ICLARM's involvement in aquaculture research and extension in
Bangladesh as a case study. With the government of Bangladesh and several

NGOs, ICLARM was at that time seeking to develop sets of institutional linkages or 'partnerships' of the types analysed in the ODI research project.

The objectives of this joint research project, which is referred to in this chapter as the ODI research, were twofold:

(1) The primary objective was to suggest institutional arrangements through which mechanisms could be created and sustained to promote effective aquaculture research and extension;
(2) The secondary objective was to document ICLARM's inter-agency aquaculture project in order to draw general lessons and produce guidelines to assist with future project planning.

Before moving on to discuss the project and the associated research, it is first necessary to make some introductory remarks about aquaculture in Bangladesh. Despite impressive increases in agricultural production in recent years such that the country is now approaching foodgrain self-sufficiency, a still increasing population has placed enormous pressure on natural resources. In the absence of new cultivable land and with only limited opportunities to further intensify agricultural production, aquaculture has become an important development strategy because it is widely perceived that Bangladesh contains a wealth of un- or under-utilised water resources (Lewis *et al.* 1996).[2]

Although fish is central to the diet of most Bangladeshis, the decreasing availability of wild fish resources in rivers and floodplains has focused attention on the potential of village ponds and roadside ditches as possibly under-utilised resources for aquaculture. Although some large landowners have traditionally undertaken extensive fish rearing practices in the countryside, more intensive aquaculture practices are new to many poorer farmers. The government's strategy for aquaculture is primarily production oriented, while most of the NGOs favour the promotion of aquaculture as a potential income generation activity for the landless and marginal households. In the promotion of aquaculture, as with much of Bangladesh's development effort, the role of external resources and foreign donors is very pronounced.

The ICLARM project

ICLARM has in recent years been engaged in a consecutive, linked series of short aquaculture projects with the government of Bangladesh, predominantly funded by USAID. The current project seeks to develop and provide low cost, low input aquaculture technologies mainly in the form of an extension message detailing appropriate pond management techniques including fish stocking densities, feeding regimes, pond preparation and appropriate species mixes which can be readily used by low income rural households towards their income generation activity portfolios. A key structural component of the project is a complex framework of institutional partnerships between government agencies, NGOs and international researchers. Training is being provided to government extension workers and NGO field staff who pass on the training to farmer groups who are seeking to pursue aquaculture as part of a range of income generation activities supported by credit provided by NGOs. Women who own little or no farm land form the 'target group' of many Bangladeshi NGOs.

The origins of the ICLARM project can be found in informal links between a number of field staff from the NGO BRAC (Bangladesh Rural Advancement Committee) and members of the Fisheries Research Institute (FRI) which emerged during the late 1980s when BRAC was starting its aquaculture programme and required some technical assistance. ICLARM was already in contact with FRI. At the same time, USAID was looking for ways to improve the effectiveness of its work in strengthening national agricultural research institutes (including FRI) in Bangladesh. A workshop was held in 1991 between all these agencies and thirty-one NGOs and the idea to work towards involving NGOs in the wider aquaculture extension effort in Bangladesh was born.[3] What resulted was an inter-agency project entitled 'Technology Transfer and Feedback Through NGOs' which was funded by USAID. Phase I ran from 1992–3 and Phase II continued from 1993–4, although this phase was later extended owing to delays until 1995. This project, which we shall term the ICLARM project, is the subject of the present chapter.

The ICLARM project involves government agencies and Bangladeshi NGOs and is designed to strengthen FRI's aquaculture research capacity and responsiveness to farmer needs along with the capacity of the wider extension system which now encompasses both government and NGOs as extension agents. There are three different government agencies taking part in the ICLARM project. The Fisheries Research Institute (FRI) is the public sector research body responsible for aquaculture and is based in Mymensingh. FRI is a comparatively new research institution without access to adequate resources and with relatively low staff morale, but was judged by ICLARM to have the potential to make a contribution to developing relevant technologies if it is provided with suitable financial support from USAID and 'technical backstopping' by ICLARM.

The Department of Fisheries (DoF) manages the national country-wide extension service but it too lacks sufficient personnel and resources, with only one Fisheries Extension Officer in each Thana, the local government unit which in some areas may contain around a quarter of a million people. The project seeks to bring NGO field workers into a collaborative relationship with DoF staff, although the DoF is driven more by production targets than by a poverty focus. Finally, the Bangladesh Agricultural Research Council (BARC) is the apex body which coordinates research and evaluates the project, although in practice BARC appears to lack a clear function within the project because it has only limited capacity to monitor activities in the field.

For the past decade many of Bangladesh's NGOs have been involved in promoting aquaculture among their organised groups of landless and marginal farmers by providing credit and technical support. Along with the three government agencies described above there are five Bangladeshi NGOs involved in the ICLARM project. In order to overcome the constraints of the government extension system the project has invited NGOs to act as additional extension agents, working in partnership with the DoF, to distribute the technology to their own target groups (usually landless farmers with an emphasis on women). The NGOs are also invited to provide feedback on adoption results and research needs to the scientists and trainers at FRI. NGO field staff are trained by FRI and ICLARM personnel alongside DoF so that this training can then be passed on to the farmers by further demonstration and training. In addition, the NGOs provide credit to their group members. The NGOs involved are BRAC and Proshika (which are the two largest national NGOs in the country), and three smaller local NGOs: Banchte Shekha and Jagorani Chakra which are based in Jessore in the south of the country and Thengamara Mohila Sabuj Sangha (TMSS) which is active in Bogra in the north.

For the ICLARM project, the development of collaborative linkages between government researchers and NGOs is central to the idea of developing low input and low cost aquaculture practices which can then be adopted and sustained by low income rural people, who can also provide feedback through the NGO field staff and DoF extension workers to the scientists. Each NGO group maintains a detailed pond data book for this purpose. The key assumptions are that NGOs need technical assistance in their aquaculture programmes, which can be met by specialised training, and that NGOs have comparative strengths in developing links at the grass-roots level (Gupta and Shah 1992).

Theoretical issues

Recent theoretical work in the social sciences has explored the different kinds of knowledge and forms of representation embodied in development projects. Drawing on Foucault's (1971) theoretical perspective on the ways in which knowledge is historically, politically and socially constructed as 'discourse', the anthropologist James Ferguson (1990) showed how a World Bank funded livestock project in Lesotho first needed to construct and represent a rural development 'problem' which could then be 'solved' by a project intervention. However, Ferguson argued that this constructed reality, or discourse, which was needed to justify having the project, did not necessarily correspond with local realities and indeed was instead driven by the wider structures of institutional power in which external agencies were engaged in interventions in Lesotho's economic and social life.

This perspective is useful for two reasons. First, it opens up to us the possibility that there is far more taking place in development projects (which may of course either assist or constrain their official objectives) than is normally described in the official project literature. If these insights can be rendered more 'visible' to project actors, it is possible that more might be learned about project progress and potential. Second, it suggests that multiple realities can be expected to co-exist within a project among the different participating agency actors, acknowledgement of which may help to explain the different motivations for becoming involved in projects and the likelihood that there will be unintended outcomes.

All this indicates that the conventional and still common view of projects as linear, controlled systems misses fundamental aspects of their nature, origins and character and that more information can be uncovered which might promote a higher level of institutional learning. In particular, efforts to understand ongoing efforts to negotiate different interests and reconcile conflicts may offer the key to improved performance. In a recent article Long writes

> The interactions between government or outside agencies involved in implementing particular development programmes and the so-called recipients or farming population cannot be adequately understood through the use of generalised conceptions such as 'state–peasant relations' or by resorting to normative concepts such as 'local participation'. These interactions must be analysed as part of the ongoing processes of negotiation, adaptation and transfer of meaning that take place between the specific actors concerned.
>
> (Long 1996: 57–8)

Long is making the case here for a methodology which he calls 'interface analysis', but his comments are also relevant to the 'process' view which we have adopted in this research which is discussed later in this chapter.

The discourse of 'partnership'

It is only relatively recently that governments and donors have 'discovered' NGOs and brought them into more prominent roles within development projects (Edwards and Hulme 1995). The government of Bangladesh has been explicitly committed to working with NGOs as 'development partners' since the Fifth Five Year Plan which was drawn up in 1990.

However, the language of partnership is a flexible one and as we have seen it can also be viewed as a Foucauldian discourse produced by prevailing configurations of institutional power and influence as development agencies, both government, NGO and International Agricultural Research Centre (IARC) compete for resources and status in relation to external resource provision. Bangladesh is one of the most aid-dependent countries in the world, with foreign assistance making up almost 8 per cent of GDP. What this means is that references to partnerships, linkages and other collaborative arrangements may not be as straightforward as they seem since they are likely to be linked to the wider resource negotiations among agency actors. For example, Biggs and Neame (1995) argue that linear models of development tend to obscure the fact that NGOs are not individual agencies but operate in a wider context based around negotiations with wider formal and informal networks and coalitions with other agencies. The negotiation *process* can be used by NGOs (and other agencies) to challenge the perceptions of donors and government and, of course, vice versa.

Two examples drawn from the ICLARM project are relevant here:

(1) The agency motives for becoming involved in partnership (in terms of what each may want to get out of the relationship) may well differ between the participating agencies. For example, while ICLARM views NGOs as carrying out the role of extending aquaculture technologies to the farmers and providing feedback, the NGO Proshika has agendas of its own, such as seeking to influence farmers and researchers towards more organic aquaculture technologies.
(2) Arrangements in practice may differ from those described in the project literature. For example, while the DoF has the mandate for aquaculture extension across the country, in practice it does not have the staff to perform this role, but does not necessarily want to be seen to delegate this task to NGOs because they are competing for similar scarce resources and legitimacy.

There are also contested assumptions behind the centrality of FRI to ICLARM's work in Bangladesh. Although this relationship clearly has its roots in ICLARM's mandate as an international research organisation to make links with and try to strengthen the 'appropriate' national research institution concerned with aquaculture research, two problems emerge with such a strategy.

The first relates to doubts in some quarters over the effectiveness of FRI as a research institution and over its operating style, both in terms of prevailing resource scarcity and institutional culture which make the envisaged shift to farmer-centred aquaculture research unlikely.[4] The second is that further doubts exist as to the overall importance of technical constraints to aquaculture, which can be solved through scientific research through projects such as this one, as compared with the social and economic ones (Worby 1994; Lewis *et al.* 1996).

It is tempting therefore to suggest that ICLARM and FRI need each other far more for the individual institutional survival of each agency than the average low income farm household in Bangladesh needs new technology for aquaculture. FRI clearly needs a donor patron, as do many such agencies in Bangladesh. Farmers at the village level trying to get more involved in aquaculture, on the other hand, are struggling with issues such as access to secure pond rights, the timely supply of appropriate aquaculture inputs and less than adequate access to credit and markets and these are discussed in more detail below.

The more that can be uncovered about these discourses the more we can assess the practical basis for partnership and the constraints within the project which may be distorting it. The ODI research suggested that in some cases the basis for partnership linkage was misplaced, while at the same time other opportunities for complementarity between agencies and projects were occasionally missed. For example, the ODA's Northwestern Fisheries Project, which is an aquaculture research and extension project with many commonalities and possible lessons to share with the ICLARM project (it has developed links with thirteen NGOs), has no formal link with FRI and has now shifted its original objectives from production and research to the extension of existing technologies. Nor does there appear to be a particularly high level of mutual learning taking place between ICLARM and ODA in Bangladesh.

Another feature of the potentially distorting effect of the dominant agency discourse is that it becomes 'necessary' for aquaculture to be represented primarily as a technical problem (because both of the key institutional partners have a research mandate) even when there is growing research evidence and NGO experience which point to the fact that constraints on the intensification of aquaculture in Bangladesh are primarily social and economic. These constraints include the poor availability of inputs, conflicts around multiple pond use, difficulties with the secure leasing of ponds, the high level of investment risk to which low income villagers are highly averse, complexities of gender in the division of labour and profit within households, and class and patronage issues in which pond owners may reclaim their ponds once they are shown to be profitable (Worby 1994; Lewis et al. 1996). These stark realities contrast with the official picture of Bangladesh as a country dotted with hundreds of thousands of un- or under-utilised ponds with the potential for massive increases in aquaculture production.

The research methodology

The ODI research project secured funding from ODA's 'holdback' facility, began in March 1994, and was scheduled to run for two years. The basic research plan was that ODI would first document the history of the ICLARM project, with particular reference to decision-making processes, successful and unsuccessful partnership linkages, and agency expectations and perceptions of project activities. This can be likened to the idea of building an 'institutional ethnography' of the project, a term employed by Escobar (1995) to describe the detailed documentation of processes and relationships using anthropological methodologies and insights. The idea then was to develop and implement, with project participants and beneficiaries, the necessary 'course corrections' which would address perceived problems and constraints.

The original intention of the ODI research and documentation project was to hold three workshops with ICLARM project participants followed up by semi-structured group and individual interviewing. The initial workshop discussions and interviews were recorded in order to provide a 'benchmark' of assumptions, attitudes and experiences against which lessons could be debated, successes and failures acknowledged and solutions evolved. Participant observation techniques were also to be employed both around the project office and on field trips to localities where the new technologies were being introduced to farmers by NGOs, government and project staff. In addition project documentation was to be consulted, an alternative history of the project drawn up to include planned as well as unplanned outcomes and comparative discussions held with other agencies involved in aquaculture.

The concept of 'project as process' was fundamental to the study and underpinned the selection of a form of process monitoring to document expectations and activities and to plan course corrections. The methodology of process monitoring and research, a loose and evolving set of alternative approaches to conventional monitoring, differs from what has sometimes been termed the 'blueprint' view of projects, which relies upon the linear planning and design of projects often as closed systems. By contrast, process monitoring and research rests on the assumption that projects are open systems in which solutions to problems can arise through experimentation and practice rather than through design. Development is seen as a dynamic process which may be perceived in different ways by different social and institutional actors and is likely to generate important unplanned outcomes (Mosse, this volume).

The current interest in viewing projects in terms of process is to some extent paralleled by recent thinking among organisational change theorists such as Dawson, who writes:

> organisations undergoing transition should be studied 'as-it-happens' so that processes associated with change can reveal themselves over time and in context . . . . This temporal framework of change can also be used to accommodate the existence of a number of competing histories on the process of organisational transition . . . . The dominant or 'official version' of change may

often reflect the political positioning of certain key individuals or groups within an organisation, rather than serving as a true representation of the practice of transition management.

(Dawson 1994: 4)

These recent developments in anthropology and organisational studies both provided theoretical underpinnings for the research study.

Understanding the ICLARM project in terms of process

As we have seen, the ICLARM project is in reality a series of projects aimed at developing and introducing sustainable aquaculture technologies. These projects have been extended and adjusted as experiences (and available funds) have allowed. As such they may be viewed as an entry point to both understanding and approaching a range of important issues around aquaculture and inter-agency partnership more generally. The problems and unintended outcomes, we would argue here, may be of value and should therefore be documented rather than lost or omitted from project documentation. This is one of the advantages of using process documentation of this kind.

Much of the ODI project was spent discussing the original intentions of the ICLARM project and comparing these intentions with what actually worked out in practice. The partnership linkages within the project were categorised and levels of partnership were identified. Through interviews with key project staff, efforts (some of which remained unfinished, which is explained below) were made to understand how these linkages had functioned, the constraints which existed and possible ways in which constraints could be overcome.

Some of these linkages proved effective, others weak. They are classified in a preliminary way in the table below. For those linkages which were categorised as weak, possible corrective action was discussed. For example, when tensions between large and small NGOs, and with local DoF extension staff were identified, the feasibility of strengthening, through lobbying and negotiation, the Association of Development Agencies in Bangladesh's (ADAB) regional Aquaculture Forum was investigated, albeit with mixed results. However, each of these linkages and subsequent attempts at course correction helped to throw more light on the overall partnership issue.

Although the ODI project did not run long enough to reach its projected conclusion, indicative findings were emerging. The ICLARM project had achieved many of its objectives which are to provide NGOs with the opportunities to gain access to technical assistance with their aquaculture programmes, to report back adoption problems encountered by the farmers with whom they work and to begin to form ties with government agencies in aquaculture for the first time. By late 1994 a total of 3,563

Points of partnership linkage and their relative effectiveness

Linkage	Mechanism	Effectiveness
Farmers with farmers	Informal contacts	M
	Demonstration sessions	S
	Household division of labour	M
Farmers with NGO field staff	Training sessions	M
	Demonstration sessions	S
	Regular NGO group meetings	S
Farmers with DoF extension staff	Occasional visits	W
Farmers with researchers	FRI field visits	W
	Monthly meetings (via NGO feedback)	S
NGO workers and DoF extension staff	Project monthly meetings	M
	FRI training sessions	M
	Special FRI workshops	M
Large NGOs with small NGOs	Project monthly meetings	M
	ADAB Forum	M
NGO staff with FRI researchers	Special FRI workshops	M
	Monthly project meetings	S

Notes:
S = strong; M = medium or varies; W = weak.
These are indicative assessments based on interviews, limited field observation and actor perceptions.

farmers (of whom 2,029 were women) had been trained, 900 ponds had been cultivated and the technology is clearly effective where it is 'properly' applied.

In particular, the feedback loop from the farmers through NGOs to researchers has been strengthened. Modifications have been made to the original ICLARM project's uniform technology package which has now been redesigned into several options in order to take account of different agro-ecological priorities based on feedback from farmers via participating NGOs in different agro-climatic areas (ICLARM 1994). Furthermore, NGOs and government researchers are now, perhaps for the first time, talking to each other about aquaculture. On paper the stated objectives have been largely met. But once we take a view of the project as embedded in a wider system of relationships and discourses some other outputs would clearly be desirable in key areas. As one might expect there are still certain areas of weakness:

the institutional level within the different participating agencies. But we have tried to move beyond this here; unlike Escobar, who presents an ultimately pessimistic picture, we believe that the dominant discourse is not monolithic but may contain some opportunities for 'room for manoeuvre' through improving the space for negotiation and transparency (Gardner and Lewis 1996).

In more practical terms the main lesson which emerges is the need for sensitivity while conducting process documentation and research, particularly with regard to the role of the external agent or agents carrying out the monitoring. This external agent requires a wide range of skills in this area, such as the ability to build trust among all sections of the project and the various participating agencies, displaying an awareness of the often unavoidable contradictions implied by different actor perspectives and maintaining a respect for the hard work put in by many of the staff involved. In the case of the present research, we were not entirely successful in ensuring that these skills were always put into practice.

Conclusions

Despite the practical problems encountered in the research, the process documentation and research methodology which was developed during this study can be seen to have generated a range of useful insights about the inner workings of development projects, and the fields of power and discourse with which they are surrounded. On a practical level the study also provides some important clues to potential future progress around both aquaculture in Bangladesh and inter-agency partnerships more widely. While 'active' partnerships are difficult to create and maintain within a resource-dependent context such as Bangladesh and while different analyses and prescriptions for promoting aquacultural development are also in competition with each other, there may be significant areas of 'room for manoeuvre'. Some agencies are clearly getting closer to confronting some of the key issues which might generate the conditions under which economically marginal households can improve their income and nutrition.

However, the resource dependency issue does not only mean that the sustainability of agency linkages is questionable, but also calls into question the nature and the focus of the technological prescription offered by many projects of this kind in Bangladesh. This is because external resource flows may help to determine the ways in which 'problems' are constructed, just as they help to structure the form in which interventions are made. The process documentation approach used in this study helps to throw the operation of these wider forces into relief in the context of aquaculture.

There are difficult decisions ahead for development practitioners working in aquaculture in Bangladesh. Research and extension initiatives will need to become more participatory, less top-down and ideally will move beyond narrow definitions of notions of farmer feedback towards models in which farmers can actually influence research agendas rather than simply commenting on technologies which are presented to them. There is also a need for development agencies to 'grasp the nettle' of addressing important social and economic constraints in ways which transcend an invocation of the assumed NGO abilities to reach farmers within what might be termed an 'instrumentalist' perspective of government and NGO relationships. By carrying out further process documentation of the kind described here, it is hoped that the disequilibrium which is likely to be generated may hold more in the way of creativity and solutions than destructiveness and contradiction. Active partnerships between NGOs and government and IARCs may yet unlock more of this potential.

Notes

1   Centre for Voluntary Organisation, London School of Economics. This research was carried out while the author was working as a Research Associate at the Overseas Development Institute (ODI).
2   Capture fisheries, despite its potential, has received rather less attention from development agencies and researchers.
3   Dr M.V. Gupta, former Senior Aquaculture Scientist, ICLARM Dhaka, personal communication.
4   Personal communication, ODA and interviews with other agencies in Bangladesh.
5   See Noble (1995) for a description of this ADAB initiative and the problems encountered in NGO–NGO cooperation in aquaculture. Noble points out that there is as yet no formal collaborative project between NGOs underway in aquaculture. An exception to this general lack of partnership is Caritas, which does give informal technical support to small local NGOs.
6   However, there are experiences which point to the fact that once farmers are convinced of the value of a technology they need little encouragement from extension workers. For example the success of ODA/CARE's rice-fish culture promotion prompts Gregory and Kamp (1996: 21–2) to write that 'a technology really worth extending is not difficult to extend'.
7   A persistent issue was the GoB sensitivities around procedure and control in dealing with outside agencies. A key weakness of the ODI research was that it had not been included as part of the original official ICLARM project proposal but was an adjunct which did not fit into a clear bureaucratic category.

References

Biggs, S. and Neame, A. (1995) 'Negotiating room for manoeuvre: reflections concerning NGO autonomy and accountability within the new policy agenda', in M. Edwards and D. Hulme (eds), Non-Governmental Organizations – Performance and Accountability: Beyond the Magic Bullet, London: Earthscan.
Dawson, P. (1994) Organizational Change: A Processual Approach, London: Paul Chapman.

Edwards, M. and Hulme, D. (eds) (1995) *Non-Governmental Organisations – Performance and Accountability: Beyond the Magic Bullet*, London: Earthscan.

Escobar, A. (1995) *Encountering Development: the Making and Unmaking of the Third World*, Princeton, NJ: Princeton University Press.

Farrington, J. and Bebbington, A.J. with Lewis, D. and Wellard, K. (1993) *Reluctant Partners?: Non-Governmental Organisations, the State and Sustainable Agricultural Development*, London: Routledge.

Farrington, J. and Lewis, D.J. with Satish, S. and Miclat-Teves, A. (eds) (1993) *NGOs and the State in Asia: Rethinking Roles in Sustainable Agricultural Development*, London: Routledge.

Ferguson, J. (1990) *The Anti-Politics Machine: 'Development', Depoliticization and Bureaucratic Power in Lesotho*, Cambridge: Cambridge University Press.

Foucault, M. (1971) 'The order of discourse', in R. Young (ed.), *Untying the Text: A Post-Structuralist Reader*, London: Routledge and Kegan Paul.

Gardner, K. and Lewis, D.J. (1996) *Anthropology, Development and the Post-Modern Challenge*, London: Pluto.

Gregory, R. and Kamp, K. (1996) *Aquaculture Extension in Bangladesh: Experiences from the Northwest Fisheries Extension Project 1989–92*, Dhaka: ODA BAFRU/Department of Fisheries.

Gupta, M.V. and Shah, M.S. (1992) *NGO Linkages in Developing Aquaculture as a Sustainable Farming Activity – A Case Study from Bangladesh*, paper presented at Asian Farming Systems Symposium, Colombo.

ICLARM (1994) *Proceedings of the Workshop on Technology Transfer Through NGOs and Feedback to Research*, Workshop notes, 20–1 March.

Lewis, D.J., Wood, G.D. and Gregory, R. (1996) *Trading the Silver Seed: Local Knowledge and Market Moralities in Aquacultural Development*, London: Intermediate Technology Publications; Dhaka: University Press Limited.

Lewis, D. and Ehsan, K. (1996) *Guidelines towards Promoting 'Active' Inter-agency Partnerships in Sustainable Development: Lessons from an Aquaculture Project in Bangladesh*, London: ODI/ICLARM Report to Overseas Development Administration.

Long, N. (1996) 'Globalization and localization: new challenges to rural research', in Henrietta L. Moore (ed.), *The Future of Anthropological Knowledge*, London: Routledge.

Noble, F. (1995) 'NGOs and the Aquaculture and Fisheries Forum of ADAB', in W.A. Shah (ed.), *Pond Fisheries in Bangladesh*, Dhaka: Environment and Resources Analysis Centre.

Worby, E. (1994) *Hitting Hairs and Splitting Targets: Anthropological Perspectives on Fish Culture Technology Transfer Through NGOs in Bangladesh*, Rockefeller Foundation conference paper, Addis Ababa, mimeo.

7

A DONOR'S PERSPECTIVE AND EXPERIENCE OF PROCESS AND PROCESS MONITORING

Ruth Alsop[1]

Introduction

This chapter focuses on the expectations and experiences of process monitoring and a process intervention involving a wide and diverse range of people and organisations – both inside and outside of government services. The programme, supported by the Ford Foundation, sought to bring about improvements in the productivity of rainfed farming in the State of Rajasthan in India. It was originally premised on the belief that multiple-stakeholder interventions, particularly under the diverse and complex social, economic and physical conditions of rainfed areas, imply a participatory and largely interactive process, as opposed to a conventional 'projectised' approach to intervention. It was also believed that, unless information was available to all actors about the 'process', it would be difficult to create the common knowledge which could form the basis of joint action and aid learning and response as events unfolded.

For the 'actor-centred' kind of development (Long and Long 1992), which both this and Chapter 8 describe, to take place in Rajasthan several things need to occur: one is that there has to be an 'enabling environment' involving policy action and high-level support. A second is restructuring of some of the organisations involved, to make them more amenable to a changed operational style. A third is investment in the human capital which both staffs and is served by those organisations. And finally, organisational mechanisms to host decentralised and multi-stakeholder interaction and decision making need to be created. While the programme described in this chapter has sought to address these needs as a whole, it is in the context of the last that process monitoring, as an enabling factor, is discussed. Process monitoring itself is only one of the elements which frame and inform interactive relationships – the information it generates is unlikely to be of practical value unless it has mechanisms developed which enable its use. In this chapter therefore process monitoring is discussed as it relates to both the broader context and other enabling mechanisms which have developed in the programme.

Programme background

The concepts and ramifications of process approaches are discussed in Chapter 1. Here, those ideas are placed in the specific and applied setting of a programme which has concentrated geographically on the Indian State of Rajasthan focusing on Udaipur District, and more recently the neighbouring districts of Rajsaman and Chittogargh (see also Farrington *et al.*, this volume).

In Rajasthan poverty is commonplace and is concentrated in those areas of rainfed farming which constitute 75 per cent of the cultivated area. In 1996, 70 per cent of the working population were engaged in agriculture and allied activities but 50 per cent of farmers had access to only 10 per cent of cultivated land. The contribution of agriculture to state GDP fell from 50 per cent in 1980 to 40 per cent in 1996 (World Bank 1997). What growth there has been in production is mainly attributable to increases in irrigated area (Kerr 1996) which is the province of wealthier producers. While the rural poor thus continue to have agriculture as the dominant element of their livelihood portfolio, mainstream programmes have as yet done little to address those issues arising from a cultural context which militates against the farmers' voice being heard by research and extension professionals.

The initiative in Rajasthan began at a time when the policy environment was undergoing a major shift and, at the level of statements, was increasingly supportive of efforts in rainfed areas and the inclusion of NGOs in government programmes. Changes in the language of agricultural research and extension projects (see Mehta 1996; World Bank 1992) also indicated a conceptual shift to demand-driven and more decentralised research and extension. However, ground truthing reveals that practice has often fallen short of expectations, particularly for the poor living in rainfed locations. Moreover, project or programme monitoring and evaluation results which often indicate poor uptake of technologies or changes in management practices have rarely fed into practical or strategic responses within existing programmes, and often only minimally into the design of new interventions. Perhaps most damaging to opportunities to learn from experience is the tendency of conventional monitoring and evaluation systems to focus on inputs and outputs. A paucity of information on the dynamics or cause and effect relations of intervention limits the understanding of why such poor returns are common or what can practically be done about it.

While the long-term goal of the programme initiative described here was to increase the contribution of rainfed agricultural production to households' economies, the shorter-term objectives have been to improve the effectiveness of relations between agricultural research and extension services and their clients, particularly poorer clients. In the attempt to overcome some of the organisational and institutional problems which hinder the progress of change for rainfed farmers, this multi-stakeholder initiative embarked on a challenging and possibly confrontational path. Two factors in particular contributed to the confrontational nature of what was being attempted. First, a 'process' which would enable change to occur in an experimental and evolutionary manner would need to have unimpeded transmission of information about events and actors made accessible to all participants. This would allow both the rationale for change to be understood and articulated by those it affected, and responses to change to be defined as events unfolded. Communication and flows of information were thus critical in the generation of the shared knowledge which would underpin action. Process monitoring was perceived as the activity which would increase the collective knowledge of both context and action. As is discussed by Mosse (this volume), making private knowledge public is problematic. It is something which both undermines the use of private knowledge as a tool of power and, where people are not used to constructive analysis or comment, can serve to embarrass or change perceptions of individuals or organisations.

Second, this transformation in communication would change traditions and forms of interaction between professional research and extension personnel and farmers. In the absence of institutionalised approval, these behavioural changes could imply a loss of prestige and authority to many of those people in the Indian cultural context. Between these two factors the status of 'establishment professionals' could be undermined – an important consideration in Indian culture.

Improving communication and relationships between poor farmers and research and extension personnel required organisational mechanisms and skills which were lacking on both sides.[2] NGOs, already in situ and with a generally good record of working with communities, were obvious contenders for a time-bound intermediary role. Thus interaction through information exchange, debate and action, between the government and non-government agencies (NGOs) became a central part of the strategy. This has been important in both strengthening farmers' capacity to articulate demand, and assisting government functionaries to respond. The strategy of this programme hinged on efforts to engage NGOs' participatory, communication and organisational skills and build on the government's technical capabilities, geographic coverage and funds. In addition, there has been explicit avoidance of replication of services or establishment of structures paralleling those which already exist.

During the time that the programme has been active there has also been a determined effort to expand the government's perception of how to work with NGOs and the subsequent tendency to simply contract NGOs for their services. While in the agricultural sector a contracting strategy has in the past yielded tangible benefits for some rural inhabitants, serious questions of scale, efficiency, technical capability and power can be raised. First, many NGOs tend to be small organisations working in a restricted geographical area. This limits the spread of new ideas, approaches and improvements in productivity. Second, there are many occasions when the development administration has staff in place responsible for precisely those activities which NGOs are contracted to undertake. Certain conditions militate against sharp divisions of labour, but the scale at which the duplication of manpower and services often occurs is inefficient. Third, many NGOs do not have the scientific or technical skills necessary for certain aspects of agricultural development. Finally, NGOs are concerned that contract relations lead to a lack of control over the design and implementation of procedures and activities. Their staff indicate a strong preference for collaborative arrangements which lead to cooperation and collegiality with government staff but which do not threaten the autonomy of either party.

The development of collegiate collaborative relations has involved a process of coalition building among those users and providers who recognise the value of, and are committed to arriving at, a common agenda for service provision within the agricultural sector. In addition to government and NGO activities, this initiative has also received inputs and support from external agencies, specifically for training and initiating process monitoring. This experimental method of operating and the processes of collegiate collaboration, particularly in agricultural development, were unfamiliar to all the actors. On-line learning and response were therefore crucial during the formative stages of interaction and negotiation as concrete activities began to take shape.

Evolutionary development and responding to on-line information is as difficult for donors as it is for many others involved in process programmes. Operationally it begs a change in the way that donors and government have traditionally performed. Donors, operating in process mode, cannot expect to 'manage', in a conventional sense, relations or activities of actors in the way that a donor working in 'project' mode could. A different set of skills and communication processes is called upon. Moreover, the donor is not simply a source of funds or directives, it has to become a partner in debate and decision. In place of being an administrator, it needs to become a facilitator of interaction and vision development. Additionally, funding cannot be simple and uniform. Process approaches require opportunistic, responsive and multi-purpose financing. For example, grants made by the Ford Foundation in this initiative have supported one or several of the following components: collaborative field action, studies, meetings (state, district and international), process monitoring, short- and long-term training and follow-up, organisational development, and documentation. Organisations supported chose what they wanted funds for and retained their autonomy at the same time as recognising the interest they shared over some issues with other groups.

Another difference in the demands of process projects relates to the importance of ensuring that senior policy makers and bureaucrats are not only informed of events, but also participate in debate and decision making. It is also critical to bring district-level staff into the same processes. To do this, fora and information flows, independent of their traditional roles and free from the laws of hierarchy, need to be established and made equally accessible to all. A donor, free from the obligations of local and national social capital, can transgress traditional boundaries and gain access to actors of all social and bureaucratic strata. In Rajasthan this ability has proved important in initially opening channels between levels of government hierarchy as well as between government and non-government agents. While it has not been possible to shake off the bonds of tradition completely, there are signs that investment in appropriate organisations and mechanisms – especially those such as process monitoring and joint fora which de-personalise relations and enable collective recognition and responsibility – is an efficient way of channelling changing human capital.

From a donor's perspective acceptance of an evolutionary approach implies a shift in both the management of change (from blueprint to process) and the apportioning of resources (from primarily technical and physical to more equally human and organisational investment). If governments are to adopt a similar approach the implications are the same. However, currently both donor and government procedures are usually driven by expenditure targets that have to be met within financial years. These need to become considerably more flexible if they are to meet process requirements. Increased support for process interventions also requires an enhanced capability among donors to understand how processes are evolving, and to make adjustments to the specifications or levels/flows of finance attached to funding agreements. Process monitoring is one activity which can assist understanding. Another, suggested above, is deeper involvement by the donor staff in the activity being financed. Both requirements run counter to current pressures to reduce volumes of aid flows and the size of aid administrations relative to each action financed.[3]

There are also pressures on governments to down-size of the public sector. Contracting out and devolution of development and democratic responsibilities to local bodies, such as (in India) the Panchyat Raj institutions may reduce staff numbers at the lowest levels. However, the changing role of the remaining government staff demands substantial investment in developing the managerial, rather than purely technical or administrative, skills of the middle and higher levels. It also requires giving attention to the political realities of reducing public sector employment. At present the union of agricultural supervisors in Rajasthan, with the support of top politicians, is in dispute with the government over just such an effort.

Building a process

The first moves in this programme were made in 1992 when informal and separate discussions were held by the Ford Foundation Programme Officer with senior government officials from the state and district; with NGO representatives from large and small organisations; and with farmers. There was consistency of response and commentary about the problem from these groups but much diversity in opinions of what, at a practical level, to do about it. While there was general agreement on the need for participatory development, debate on the potential of collaboration between NGOs and government indicated very different perceptions of possible working relationships, particularly in relation to decision-making power, accountability and fiscal responsibility. Despite these differences of perception over the form that new working relations might take, government and NGO staff expressed interest in further interaction and met formally to discuss strategic options at a state-level meeting supported by the government but organised and funded by the donor.

During the first year of discussion a small number of grants had been made to NGOs of different sizes and with different capabilities who were willing to try to bring farmers and government staff closer together in villages. Initially these actions were not so important for what they achieved physically as for the focus they gave to the debate on the practical matters of what to work on and how to work together. They also gave NGOs who did not need or want donor support the opportunity to observe and discuss the opportunities and cost-effectiveness for developing the primary livelihood sector of agriculture.[4] All this was documented informally by the donor through circulated letters and notes which reported and commented on visits, discussions and issues.

In the early days there was no prescribed structure or pattern of activities. While there was agreement on the primary goals and objectives, there were no 'project' boundaries, no script and no limits on who participated. Three factors determined this. The principal one was that it was considered essential to a stakeholder process to let those stakeholders develop and define the activities and relationships which they thought appropriate. The secondary reason was that a more structured or pre-determined approach would have needed more informed actors. The paucity of experience of these types of process in the agricultural sector of Rajasthan meant that there was no reference point or history on which to base ideas of how to formulate new action. Finally, even if there had been the experience to draw upon, unlike bi- or multi-lateral donors, the Ford Foundation was not a donor which could finance exploratory planning processes or field experimentation on a large enough scale or which, in the final stages of project definition, was in the business of negotiating structural change within government. It was antici-pated that the evidence, emerging from attempts of clients to interface with government services, would both provide evidence of the need for structural change and offer ideas of how this could be done.

To maintain momentum the donor organised a further formal state-level meeting towards the end of the first year. NGOs and government reported on the collaborative activities in which they had been involved over the past months and the Ford Foundation Programme Officer put forward a working strategy paper. This paper was intended as a thought piece and represented an attempt to offer actors, not a 'blueprint', but a framework on which to hang the different activities which were beginning to take shape.

During this second state-level meeting and in smaller ones that had taken place over the past year it was clear that, while oral information relating to the organisational, procedural and behavioural problems of working together might be shared informally, formal presentations and reports on activities repeatedly skirted these critical issues. Several reasons, which needed addressing if collective action was to succeed, underlay this:

- direct individual and public reporting can personalise issues and jeopar-dise relations;
- these were not issues or topics which the actors were used to considering or felt that articulation would change;
- no locally controlled or institutionalised mechanisms for inter-organisa-tional information exchange, debate or decision making existed at that point in time.

A support system for the emerging coalition was needed.[5] It was apparent that joint activities – whether collective or bilateral – could not be effectively undertaken, understood or modified unless channels for monitoring, infor-mation exchange and decision making were developed. If issues of process were not shared and understood, mistakes would be repeated and lessons lost. A process monitoring system focusing on information flow and use was therefore essential. If there was no information flow shared knowledge would be uneven, understanding of events and opportunities disparate and the likelihood of achieving the degree of consensus necessary for collective action very limited. In addition, to use information effectively and collec-tively there needed to be fora for discussion and decision making.

While several of the participating organisations, including the Ford Foundation, already had some form of monitoring system in place it was rare for these to focus on process, particularly the organisational or institu-tional dimensions. Most were concerned with outputs and products rather than how and why particular points or achievements were reached. Additionally, in Rajasthan it was rare to find a system which used information for decentralised or on-line, rather than end-of-project/phase, decision making. Where this occurred systems were usually very informal; only useful in small organisations; and produced information which was not robust enough to withstand external scrutiny. The only system for inter-organisation exchange was ad hoc in the extreme; prone to information distortion; and not usable for collective decision making.

A search for a local agency able and willing to initiate, systematise and develop local capacity for both intra- and inter-organisational process monitoring went unrewarded. The Ford Foundation thus opened discussions with ODI. A grant was made by ODI on the basis of previous experience with research on NGO–GO collaboration. Their presence was first felt in Rajasthan at the second state-level meeting. Attending this meeting were government representatives from District and State, NGOs and representatives of several donors.

Monitoring, information sharing and decision making

ODI was confronted with a task which, in the beginning, few actors understood or appreciated as a non-judgmental and constructive part of the process they had embarked upon.[6] The donor, however, had clear expectations of process monitoring. At its broadest process monitoring would:

- ensure multi-stakeholder learning which would lead to more effective collaborative activity;
- enhance individual organisation's capacity to assimilate knowledge and inform action;
- assist in the institutionalisation of learning organisations capable of evolution;
- monitor the Foundation's own operational efficiency.

Conceptually for the donor, the process monitoring vision was of discrete but nested systems, each owned and managed by an independent agency but each selectively contributing to and accessing a common information stream (Alsop 1996). This demanded recognition that the need for process monitoring varied according to the actor – each having its own agenda and capacity to gather, absorb and use information. Such a system also depends structurally on 'nodal points' for analysis and re-presentation and 'niches' or platforms for multi-stakeholder reaction and decision making. However, although the system existed as a donor concept, on the ground it had to evolve as those using and contributing recognised process monitoring costs and benefits and bought into devising appropriate and manageable systems.

To date, the early concept remains surprisingly appropriate – although the methodology and techniques used for information handling are somewhat different from the original vision. Systems in Rajasthan are still in a state of 'becoming' but there is enough evidence to show that a number of organisations are committed and beginning to develop their own (discrete) process monitoring systems. As these, and their genesis, are described more fully in the next chapter, here the commentary is restricted to a level of generality which both illustrates the elements and highlights areas where differences in interpretation occur.

Each discrete or independent system is peculiar to the organisation developing and using it and each varies in both the degree of formality with which they handle information (oral/written;[7] loose/highly structured) and decision making. The tools and techniques of information handling are as varied as the actors, but use of conventional social science techniques in any pure sense is rare. No trained sociologist or anthropologist has worked with the organisations concerned. The principal objective in these independent systems is not to arrive at an objective 'reality', or to apply analytical techniques with academic rigour, but to understand from within and to use that information to make adjustments to everyday work. This role of process monitoring can be likened to what Mosse earlier refers to (Chapter 1, this volume) as 'develop[ing] agency capacity to undertake a . . . complex task'. In both intra- and inter-organisational settings it is, however, a role in which the distinctions between process monitoring and other forms of monitoring are often blurred.

While not a substitute for more conventional monitoring activities, process monitoring is in several cases being treated simply as an analytical component of monitoring as a whole. It is that part which helps understanding of why and how input (a) led to output (b) – or not, as the case may be. Because of the techniques used, which each organisation has defined themselves, in only one case has there been data overload – and this was where there was an attempt to implement a traditional and comprehensive system. As these systems are not 'scientifically' designed in the positivist sense of the word, it is possible to question the validity of some of the data. Findings are temporally and spatially specific, and tend to focus on issues or events which are perceived as locally important. However, it is worth looking at the effectiveness of the monitoring activities before rejecting the information they produce on the basis of poor design. Unlike some of the larger and more cumbersome project monitoring systems generating massive amounts of data which take both specialists and a considerable time to analyse, these smaller and less academically respectable systems do generate information that is useful and used. Pragmatically these systems are justified. However, there is also a role for more rigorous studies of specific issues which arise out of process monitoring as we see it being undertaken by many of the organisations in Rajasthan. There is also arguably a role for quasi-academic research into issues which affect programme design – such as on the nature of how local people organise for collective action and how this changes over time. Nevertheless, it is suggested that this should not be regarded as process monitoring, nor should there be any attempt to overload information systems which include process components and which are proving useful in feeding into strategic and applied action of individual organisations.

In no case yet is there an intra-organisational monitoring system containing this process element which is routine enough to become stale and unresponsive. This may of course emerge as a second generation issue. A different order of problem is the fact that process monitoring has often been queried as a valid activity. Specifically this has occurred when process monitoring activities, true to the concept of the 'learning organisation' (Burgoyne 1992; Checkland 1989; Senge *et al.* 1994) have led to a cycle of tension, resolution and change within organisations. In organisations where the purpose of process monitoring is not fully appreciated by all and where mechanisms do not exist to 'manage' differences in interpretation, this can lead to very uncomfortable situations – which in practice most people prefer to avoid. The manner in which organisations have used their information, the fora used for its discussion and the manner of decision making contingent on process information have differed with each organisation. Many factors affect this including size of the organisation, mandate, its management structure and procedures, interests and skills of staff members, and its relationships with villagers.

Although in *intra*-organisational situations process monitoring has been used as a means of 'engagement in institutional processes of negotiation and consensus building' (Mosse, Chapter 1, this volume), the experiences in Rajasthan suggest (to the donor) a slightly different interpretation of the role of process monitors in *inter*-organisational situations. While in Rajasthan the inter-organisational process monitors have engaged directly in the process itself, the explanation for this relates more to the specific context (including the need for those new to the programme and place to understand context, legitimise expatriate involvement and justify the act of process monitoring) and personal skills and proclivities of those concerned, than to a principle of information gathering and use. The need for those taking information into the public domain through process monitoring to be uninvolved in interpretation is made clear as a local agency in Rajasthan assumes responsibility for process monitoring. It is apparent that unless this organisation (assuming the role of a supportive 'nodal' agent to the coalition) can remain impartial in the way it handles information, it will be accused of dominant and interpretative behaviour – something which would be detrimental to the functioning of the coalition.

The role of information in inter-organisational situations dependent on informal participation in a coalition (where information is made public) is very different from its use within an organisation which has formal membership and boundaries which differentiate insiders from outsiders (where information remains private to that organisation). In an established organisation the stakes that staff have in that organisation's survival are higher than those which participants have in the survival of a coalition of multiple organisations. Coalitions are fragile, often transient, and dependent on a rule of equality in engagement in all activities – including information exchange. Information transmission among diverse actors sharing an interest in a specific set of issues or actions has a particular function – it not only makes fragmented or individual knowledge public, it is also used to create collective knowledge.

The formation of collective knowledge is dependent on information being made available and on it being assimilated, and re-expressed as a collective understanding. The belief that knowledge has to be common in multi-stakeholder situations is rooted in the idea articulated by Foucault (1971) that knowledge is power. Therefore for a coalition to function effectively each member has to be equipped with the same knowledge of the thing being discussed or about which decisions are being made. This avoids the possibility of actors using knowledge in either the behavioural (using knowledge to their own advantage at the cost of others) or structural sense (to reinforce their position) (White 1993). Either may undermine or cause the breakdown of coalitions. Information will not become a public good, however, unless there are determined and explicit efforts to make it so.

The way in which information is presented and made available for collective assimilation and discussion thus deeply affects the way that shared knowledge is generated. In a coalition such as that in Rajasthan, each actor will interpret not only according to what is already known but also the way in which he or she knows. This has been a common problem in development interventions where there are 'structural discontinuities' (Scoones and Thompson, 1993) in the epistemologies of different actors. For collective action to occur there has to be some degree of consensus and that is dependent on common understanding of what is being dealt with and what should be done about it. Information contributing to the knowledge base thus has to be attributed and recognised as coming from a particular source and not as common knowledge until it has been shared, often contested and re-expressed collectively in its original or modified form.

Currently, a key output of process monitoring for the coalition as a whole is a bi-monthly publication – *Recent Developments*. This publication is mainly dependent on attributed contributions from coalition participants, and thus is seen to offer a level playing field to those who wish to share information irrespective of their organisation or position. To date there has been no substantive editing, and as yet nothing has been submitted which has been considered too contentious or misrepresentative. One issue carried an editorial, but this has now been discontinued. If the organisation which handles this publication – which should be regarded as a collectively owned product – were to use it to pursue an agenda or to engage in 'advocacy, facilitation or nurturing' (Mosse, Chapter 1, this volume) without the approval of coalition members it would threaten a process dependent on collective responsibility and sanction.

Similarly, a comparable concern of acting in a service, rather than a pro-active, role applies to the way the same organisation oversees one of the 'niches' – the NGO–GO Forum – for discussion and to an extent, collective decision making. The Forum is interesting not only as a manifestation of the coalition, but also because it links both documented and oral process information to discussion and decision making. Issues emerging from both are discussed and occasionally decisions made to take action. These have either related to collective action, as for example in the case of joint field visits and the financing of an agricultural research fund, or have been of a commissioning nature, such as in asking one or more people to take action. The latter has included letter writing to represent the collective view, studies on issues of interest, and the formation of a working group on NGO–GO collaborative strategy.

The Forum emerged as a purely local initiative and was not a pre-requisite (although welcomed) part of a donor programme. Because of the way it emerged it is regarded as a locally owned expression of interaction between agencies. However, the organisation which hosts it and records the minutes keeps the meetings open to all and avoids directing proceedings. If its behaviour was otherwise the Forum would become perceived as 'owned' by, and the responsibility of, that organisation, not of the coalition. The emphasis placed here on the need for impartiality in acting as a node of information management or as a host of a platform for discussion does not deny the difficulty of this role. Neither does it fail to recognise that on occasion inter-organisational process monitors, because of their proximity and knowledge, may become, of their own volition or by request, more than suppliers of information. In practice roles rarely have sharply defined boundaries and any person engaging in any way in a process will affect it to some degree. However, it is suggested that the norm of process monitors in intra-organisational situations should verge on the side of the observer rather than the activist.

Summary

The experience in Rajasthan indicates that for process approaches involving diverse and multiple stakeholders to be effective there must be both enabling environments (policies and programmes) and enabling mechanisms (channels for information flow and convergence, fora for debate, and platforms for decision making). These are essential if the interests of the multiple actors are to become effectively shared, or coalesce, to the point of bringing about change. The process has been one in which coalitions of interest, as loosely structured and ungoverned ways of organising people, appear as useful ways of conceptualising the organisation of actors in such types of interventions. These coalitions, though, need to be locally owned and the 'enabling mechanisms' must be locally developed.

For the donor a primary focus on organisational and institutional issues has meant departing from traditional project cycles and withstanding pressure to prove immediate impact. It has entailed operating in a facilitative and responsive manner – both professionally and in terms of the activities financed. The positive elements of this participatory process include greater local ownership of ideas and action than would predictably occur in a pre-defined project, strong identification with the ideas and issues by senior officials at the state and district level, organisational mechanisms managed and owned without external direction which implies post-intervention sustainability, and joint action between NGOs and GOs which has occurred independently of external interest.

However, the emphasis placed on stakeholder definition of action has led to problems associated with actors having no 'structure' to work within and no 'outputs' defined for them. These problems have been manifested in: a very slow pace of change in farmers' fields; long lead times for deliberation of ideas and emergence of action; vulnerable or weak systems of accountability among actors; poor definition of roles of various actors; unsustainable requirements for external people to maintain state and senior bureaucrats' interest; and a continuing interface with government extension systems which are not able to deal with client-driven agendas.

On balance the above suggests that process approaches would benefit from being more structured. This does not imply a rigid approach in which activities, procedures and intended outputs are all predetermined, as would be the case in a traditional 'blueprint' project. Rather, it implies stakeholders planning together and agreeing in advance the preconditions for successful collaboration, the respective roles of different organisations, shared objectives and, where possible, intended concrete outputs. These agreements relate to *principles* of interaction, rather than mapped actions. This level of structuring is considered appropriate to increasing speed and efficiency in reaching goals without imposing specificities which would constrain responsive action. The next few years in this programme, which after withdrawal of the Ford Foundation from financing in the agricultural sector is being supported by a bilateral donor for a year of stakeholder planning in an expanded geographic area, will provide the material to test this hypothesis.

The original donor view of process monitoring as an aspect of a process approach has again been altered and coloured by experience. To a degree process monitoring has created better information flows which have enabled response to needs as the process unfolds – this is perhaps more true for individual organisations than for the coalition as a whole. Process monitoring is becoming adopted voluntarily as an internal tool of learning by some organisations and inter-organisational process monitoring activities and outputs are locally managed and perceived as useful by stakeholders. The approaches are less rigorous than the donor originally envisaged but the outputs are being used in a way which a more systematic approach to information gathering and analysis might not have encouraged.

After two years of support from an outside agency (ODI), a local 'nodal' agent now takes responsibility for inter-institutional process monitoring. The techniques introduced and used by ODI have been broadly, if not strictly, anthropological – dependent on discussion, interviews and participation in field visits and meetings. Stimulation of interest in process monitoring and provision of information for immediate use by coalition members has taken precedence over social science analysis. There has been no attempt to undertake longitudinal or academically acceptable studies of cause and effect relations over time and no use made of more conventional techniques of information gathering. While this approach has, from an academic and project planner's position, hampered understanding of how to support process interventions, it has enabled local actors to identify with the immediate use of process information.

As Mosse comments in Chapter 2, generally for the external process documentors 'process monitoring amounted to providing a "communication service" to address local concerns and to resolve immediate difficulties'. This has been a useful service but some of the information needs, in particular those which would inform on how to develop and manage multi-stakeholder processes, have not been met. In this way process monitoring activities have perhaps unfortunately been used mainly as a vehicle to 'validate collaboration' (Mosse, Chapter 2, this volume) rather than one which has sought to question and test the validity of collaboration.

However, process monitoring in Rajasthan has enabled and continues to enable learning and decision making at a number of levels. As part of an intervention strategy it can offer the opportunity for a donor to learn about the effectiveness of the action it is supporting, the dynamics of multi-stakeholder action, and how change is best managed structurally and strategically. Perhaps more importantly in Rajasthan it has begun to provide grantees and others involved in the initiative with a mechanism which can publicly inform, help them understand each other's position and action, and provide the material for debate and change. Additionally, it is assisting in the creation of independent learning organisations, which is essential when the business of those organisations is change. If organisations use new knowledge to institutionalise in themselves the characteristic of evolution they can become more effective vehicles for hosting the upgraded human capital in which development invests. Process monitoring, as a system of information flow and analysis, when linked to structures enabling debate and decision making, is therefore a functional component of multi-stakeholder development.

Notes

1 International Food Policy Research Institute (ex-Programme Officer Sustainable Agriculture, Ford Foundation, New Delhi).
2 'Organisation is essential to the achievement of effective agency .... It is the stabilising and fixing factor in circuits of power' (Clegg 1989: 17).
3 While Ford Foundation staff are subject to some of the same fiscal and managerial pressures as other donors, the organisation also operates in a manner allowing the fiscal and administrative flexibility central to a process strategy. The flexibility includes loose definitions of short-term goals, outputs or targets; the ability to fund opportunistically and rapidly; and positioning of an agent (in this case the Programme Officer) at the interface of the process and as a donor able and authorised to filter and respond to events and demands.
4 Few NGOs in Rajasthan had at this time focused on agricultural production. Where they had been involved in natural resource development both their technical and organisational work had concentrated in environmental restoration and forest or wasteland management. The Ford Foundation initiative was timely inasfar as much of the physical work completed as part of these earlier concerns had created the preconditions for improving agricultural output.
5 The use of coalitions as an organisational mechanism in multi-stakeholder situations is discussed in Alsop et al. (forthcoming).
6 See the contribution in this volume by Farrington et al. which describes ODI's experiences and takes the account of events further forward.
7 Particularly in smaller organisations process information is transmitted orally. Because of this the term process monitoring *documentation* is avoided.

References

Alsop, R. (1996) 'Nests, nodes and niches: a system for process monitoring, information exchange and decision making for multiple stakeholders', mimeo, IFPRI: Washington DC.

Alsop, R., Farrington, J., Gilbert, E. and Khandelwal, R. (forthcoming) *Pursuing Partnerships: Coalitions For Agricultural Change*.

Burgoyne, J. (1992) 'Creating a learning organization', *RSA Journal* CXL, 5428 (April): 321–32.

Checkland, P. (1989) 'Soft systems methodology' *Human Systems Management* 8:273–89.

Clegg, S.R. (1989) *Frameworks of Power*, London: Sage.

Foucault, M. (1971) 'The order of discourse', in R. Young (ed.), *Untying the Text: A Post-Structural Reader*, London: Routledge and Kegan Paul.

Kerr, J. (1996) 'Sustainable development of rainfed agriculture in India', *EPTD Discussion Paper*, No. 20, IFPRI: Washington DC.

Long, N. and Long, A. (eds) (1992) *Battlefields of Knowledge: the Interlocking of Theory and Practice in Social Research and Development*, London: Routledge.

Mehta, M.L. (1996). 'Agricultural Development in Rajasthan: Some Issues of Policy and Growth', *Rajasthan Economic Journal*, 14, 2 (July).

Scoones, I. and Thompson, J. (1993) 'Beyond farmer first. Rural people's knowledge, agricultural research and extension practice: towards a theoretical framework', *Sustainable Agriculture Programme Research Series*, 1 (1), London: IIED.

Senge, P., Kleiner, A., Roberts, C., Ross, R. and Smith, B. (1994) *The Fifth Discipline Fieldbook; Strategies and Tools for Building a Learning Organization*, New York: Doubleday.

White, G. (1993) 'Towards a political analysis of markets', in *IDS Bulletin* 24 (3) Institute of Development Studies, UK.

World Bank (1992) 'Staff appraisal report – India: Agricultural Development Project, Rajasthan'. Washington DC: World Bank.

—— (1997) *Reforms for Restoring Fiscal Sustainability in Rajasthan*, Country Operations, Industry and Finance Division, Washington DC: World Bank.

8

PROCESS MONITORING AND INTER-ORGANISATIONAL COLLABORATION IN INDIAN AGRICULTURE

Udaipur District and beyond

John Farrington, Elon Gilbert and Rajiv Khandelwal[1]

Introduction

Two features distinguish the experiences presented here from those of Chapters 3 to 6 in this book:

- they are located in a context of efforts by indigenous (in this case, Indian) organisations, governmental and non-governmental, to identify how they might work more closely together, a context in which funding by the donor sponsoring process monitoring (PM) – and by donors in general – plays an important facilitating, but secondary role;
- they derive not from projects, but from day-to-day interaction among different types of organisation in which projects, programmes, administrative and legal procedures and politics all play a part.

Put bluntly, whilst recipient governments and NGOs may be willing to accept PM as part of a major aid package, in the absence of such a package they are likely to see it as an external imposition on their normal work patterns unless it quickly demonstrates that it has advantages to offer, and can be introduced in ways which do not directly challenge vested interests. This chapter reviews efforts to apply PM in the latter context over almost three years in Udaipur District of Rajasthan, India. The experiences reported here permit insights into the opportunities and, importantly, limitations facing external organisations introducing PM techniques into contexts largely unrelated to donor funding, i.e. outside of traditional projects where monitoring activities are commonly directly associated with the validation of progress and the release of funds.

The external organisation which introduced PM was the UK Overseas Development Institute, a private, non-profit policy research institute, whose previous work in India had included a study of the scope for closer collaboration between NGOs and government research and extension services in technology development and dissemination. Both ODI's introduction of PM, and its earlier work on NGOs in India were funded by the Delhi office of the Ford Foundation (FF).

Not surprisingly, the interpretation of PM here differs somewhat from that in other papers. Our basic concern has been to explore with the actors (GO and NGO) directly involved how the cycle of collaborative action, documentation and reflection can be strengthened. As Chapter 7 by Alsop in this volume indicates, following extensive local consultation, the Ford Foundation took the initiative in funding certain joint actions. The initial ODI concept was of a fairly tight two year schedule of action, documentation and reflection. Following consultation with participating organisations, the ODI team introduced methods of documentation which were new to the actors concerned and were specifically intended to facilitate their monitoring of how attempted collaborative actions had performed. In a parallel initiative, ODI and a local collaborator, the Vidya Bhawan Society Krishi Vigyan Kendra (VBKVK – Farm Science Centre), with the agreement of local NGOs and GOs, established a 'KVK Forum', the purposes of which included information exchange and the design and monitoring of joint action. Thus, PM is used here as shorthand for new forms of documentation and related monitoring of inter-agency collaboration and, to a degree, intra-organisational processes and performances. However, the set of activities involving ODI in Udaipur quickly became wider than this, particularly during the third year extension of the initiative. Over the three year period, the relations among some sub-groups of actors developed faster than among others, individual actors adapted specific kinds of documentation to suit their own purposes, and a wider set of actors at state and national levels began to monitor events in Udaipur informally. Almost imperceptibly – although elements of process-related documentation and its specific monitoring remained – the conventional devices of democratic pluralism – the agenda for and minutes of meetings, newsletters and published correspondence – came to be used with more skill and confidence by the actors involved. In this context, ODI's role expanded to include support for the smaller, hitherto marginalised organisations, to engage both in PM and in the use of more conventional tools. This was achieved largely through informal networking. In the body of this chapter we describe this evolution in more detail, assessing how far the new skill and confidence are attributable to the intervention of 'outsiders'.

In terms of the concepts presented in Chapter 2, we argue that the experience presented here began as, and largely remained, an *interpretative analysis* in which outsiders exercised a degree of independence from local events in order to 'perceive different perspectives, identify communication gaps . . . and more generally to generate critical reflection on practice'. However, the role of PM here evolved over time: it quickly became apparent that an initial intention mainly to conduct an analysis *of* collaboration and assess its impact was unrealistic given the time (2 years) and resources available; as well as the sensitivities and capacities of partner agencies. The emphasis therefore shifted towards a strategic involvement *in* collaboration, in which, for instance, explicit efforts were made to increase the self-confidence and capacity of smaller organisations to make their voice count. Finally, we would argue that this approach to PM assists in the resolution of potential day-to-day conflicts over roles and responsibilities without resorting to simplistic models (such as PRA). We believe that it has made a contribution, in the words of Chapter 2, to the notion that '[i]terative planning and implementation demand "resolutory" roles which build consensus based on a better understanding of different as well as shared roles'.

The chapter is structured as follows:

- it sets out the biophysical, socio-economic and institutional contexts of agriculture – the sectoral focus of this effort – in Udaipur. It also briefly summarises the initial involvement of the FF in supporting collaboration and PM in Rajasthan;[2]
- it examines the components of PM used, initial responses to them, and modifications not only to the components but to the entire strategy in the light of these responses and of changes in the wider context;
- it draws tentative conclusions on the wider prospects of PM as a tool for understanding and furthering partnerships between government organisations and NGOs (whether service-providing or membership organisations) in 'indigenous' settings of this type. It also draws conclusions on the role of external actors.

Context

Udaipur District located in Southern Rajasthan has an estimated population of 1.5 million, 55 per cent of whom belong to tribal groups. These depend primarily on agriculture for their livelihoods, and mainly cultivate rainfed cereals, geared to meet subsistence needs. Nearly 50 per cent of all the farm families in the district cultivate less than one hectare.

Low and erratic rainfall, extremely limited surface and groundwater resources and heavy soil erosion characterise the natural resource context. The impact of a rapidly deteriorating ecological base in the area has been most severe for small and marginal farmers, who supplement low and unstable farm incomes with seasonal employment in industrial areas, primarily in neighbouring Gujarat.

Udaipur city has more than 300,000 population, and a long, rich history as the capital of the Mewar State. As the district centre, it serves as the local headquarters for government of Rajasthan (GoR) departments, including agriculture, horticulture, water resources, livestock and rural development. Rajasthan College of Agriculture, a campus of Rajasthan Agricultural University (RAU) is located in Udaipur, and RAU manages most of the network of Krishi Vigyan Kendras (KVKs), or Farm Science Centres, in the state. The KVK located on the outskirts of Udaipur in Badgaon is managed by an NGO, the Vidya Bhawan Society (VB). The VBKVK has played an increasingly important role in PM in the district.

In recent years, the State and national governments have endeavoured to improve the quality and quantity of services to poorer rural communities. The limited gains in productivity that have been achieved reflect not only the difficulties of climate and soils, but also the networks of caste, class and patronage which strongly influence the distribution of benefits from most programmes of this sort. Many poor communities have had lengthy histories of frustrating experiences with government programmes and tend to view new initiatives with suspicion.

In response to these difficulties, state and national governments have sought the involvement of private non-profit organisations. Very simply, it was felt that service-oriented NGOs could serve as effective intermediaries between government programmes and beneficiaries, identifying the needs of low-income rural families, drawing on government programmes as appropriate, organising joint action among farmers in such spheres as watershed management, and so generally generating greater benefits at the same, or lower, cost to governments.

These principles are incorporated in the Agricultural Development and Watershed Management projects, both supported by the World Bank, and they are enshrined in the Government of India Guidelines for Watershed Management of 1994, and Guidelines for Joint Forest Management of 1990. In parallel, the GoR is considering the decentralisation of agricultural research and extension activities by giving more authority to district and zonal-level staff in an effort to make services more responsive to local requirements and to facilitate multi-agency approaches.

Udaipur has long been a centre of NGO activity in the region. Among the best known are Seva Mandir and the Vidya Bhawan Society which trace their origins back several decades. In recent years, a large number of smaller NGOs have emerged, many started by individuals who earlier worked for larger NGOs. Today, more than seventy NGOs operate in the district representing a wide range of philosophies and operating styles. Most regard the conduct of government departments as part of the problem of inappropriate and ineffective service delivery, based as they widely are on rent-seeking and on social and political predispositions which perpetuate the poverty and low status of tribal communities in particular. NGOs remain sceptical that government behaviour can be fundamentally changed, but are active in pressing for specific actions which protect or otherwise help the communities in which they work. While some avoid participation in government programmes as a matter of principle, many NGOs, with some reluctance, regard the resources available through government programmes as perhaps the most promising means by which poorer communities can improve their livelihoods and sustain what remains of their natural resources. Further, most NGOs recognise that they do not have the expertise required to promote technical change in agriculture and some recognise the need to draw this expertise either from larger NGOs (e.g. SPWD and BAIF) or from government services.

Government initially sought to contract NGOs to provide specified services in agriculture, but many NGOs see this potentially as an abrogation by government of its responsibilities. It has therefore generated little response from NGOs throughout Rajasthan, and policy promulgated at the highest levels (Secretariat for Agriculture, Government of Rajasthan) now focuses on the desirability of partnerships between NGOs and government, in which each side brings its comparative advantage to bear, so making the whole greater than the sum of its parts. NGOs' experimentation with diverse approaches to agricultural extension is seen as an important learning opportunity for government in the context of such partnerships.

However, with the exception of a few enlightened government officers, there has been a marked reluctance on the part of line department staff to implement this policy. This is for five broad reasons:

- the threat that they perceive to their own job security if NGOs are to be allowed a greater role in extension;
- concerns (in some cases valid) over the financial probity of NGOs;
- the inconvenience of having to respond to demands 'from the grass-roots';
- concerns (again, in some cases justified) over the competence of NGOs to identify appropriate technical options with and for farmers;
- concerns over the levels of transparency and accountability likely to be demanded by NGO partners.

An array of external agencies complete the cast of organisations. FF has recently played a prominent entrepreneurial role in GO/NGO collaboration, as is discussed in more detail below. The World Bank and national government agencies are major sources of support for large agricultural and natural resource management programmes which increasingly feature facilities to encourage the participation of the voluntary and private sectors. The Indian Council of Agricultural Research (ICAR) has a range of programmes in which NGOs participate. These include contracting the management of a number of ICAR KVKs to local NGOs, as in the case of the VBKVK in Udaipur. The Swiss Development Corporation (SDC) through Intercooperation has been active in supporting rural and agricultural programmes among a set of NGO partners and is focusing its support largely on the voluntary sector. SDC has also recently agreed to fund a collaborative watershed development project in two districts (Chittogargh and Alwar). The Swedish International Development Agency (SIDA) is supporting an innovative collaborative natural resource management project (PAHAL project) in neighbouring Dungarpur District in which local NGOs have provided training and advisory assistance to village organisations who in turn are being given increasing responsibility for the local planning and implementation of specific activities, such as reforestation, soil and water conservation and agricultural extension. In all collaborative ventures, a major dilemma facing NGOs is whether the resources available through government and donors can be accessed without compromising the integrity of the NGO or risking significant negative effects for the communities they serve.

As early as 1992, the Ford Foundation (FF) perceived that if NGOs, the farmers' groups with which they worked, and government services were to work together in ways more profound than merely contractual relationships, then significant efforts would have to be made to create within both government and NGOs a culture of understanding how interactions functioned, and what contributed to their success or failure. Seeing PM as a means of promoting such a culture, of understanding partnerships and of facilitating any necessary changes, FF provided resources for the introduction of various forms of PM within the context of a wider set of initiatives. FF engaged ODI to introduce PM techniques in Udaipur through both expatriate and Indian staff, primarily to follow and better understand the outcomes of collaborative engagement. A separate allocation was made to the VBKVK for the establishment of an information centre. The wider initiatives supported by FF included (see Alsop, this volume): (i) strengthening of a State-level NGO liaison committee; (ii) training of government and NGO staff in PRA techniques; (iii) identifying innovative ways of using the media in extension communications; (iv) commissioning of a state-wide study by PRADAN of experiences in GO–NGO collaboration in several fields; (v) support to a number of NGOs willing to bring government services and farmers together; (vi) and the establishment of a small fund to allow NGOs and the farmer groups they represented to commission research from the public sector which would not otherwise be done. Most of these initiatives contained elements of PM, including intra-organisational forms in the case of (v) and (vi). FF also engaged in several rounds of meetings at both state and district levels in order to explore perceptions on the prospects for multi-agency partnership, some of which focused on the use of PM as tools for monitoring joint activities.

PM in Udaipur: changing techniques and strategies

ODI started field activities in Udaipur in February 1994 without a detailed PM plan. The initial visit of the ODI research associate was devoted to becoming acquainted with the organisations and individuals interested in collaboration in the district and elsewhere in the state. Substantial time was devoted to discussions of PM purposes and approaches. It was envisaged that some forms of PM might be suitable for use by the organisations themselves for internal purposes as well as assistance in better managing relations with others. A methodological statement was drafted in March 1994, primarily to serve as a basis for discussions. The statement, which was finalised a year later, focused on the collection of information on developments with a view to both informing participating organisations and assessing changes in organisational performance and impacts as a consequence of collaboration (Gilbert *et al.* 1995). This approach encountered several problems:

- despite ODI's emphasis on the concept of understanding *process*, a number of organisations, both GO and NGO, had difficulty in distinguishing between PM and conventional performance monitoring, and so were reluctant to become engaged;
- a number of (mainly) NGOs were not clear what was being monitored since they did not feel that much if anything was actually happening or had been set in motion; and
- doubts arose over whether it would be possible to produce conclusive evidence on changes in organisational performance, especially if assessments focused mainly on the exploratory and initial implementation phases of collaborative efforts. Impact assessments were regarded as particularly unrealistic given the 2 year time frame for ODI's work.

ODI and VBKVK hosted a workshop to share experiences with PM and ideas about future directions in November 1995 in Udaipur. However, most of the discussion focused on experiences with collaboration itself. Most participants had not given much thought to PM at this point and remained unconvinced of its utility. There was a general (if not unanimous) appreciation of the efforts of the ODI team members on behalf of collaboration and individual organisations, but these activities remained unconnected with PM in the minds of most.

ODI has sought to have its PM activities guided by what organisations perceived to be useful in documentation, analysis and communications, primarily (but not exclusively) related to collaboration. In the early phases, substantial time was devoted to assisting individual organisations to explore collaborative engagement. In this capacity, ODI played an *advocacy role* in promoting engagement, without which there would have been little to

138

monitor! As a few collaborative activities got under way, ODI has played more of a *nurturing role*, assisting participants with communications during the critical early stages. Those roles benefited from current knowledge of what was available and happening, and required an understanding of the organisations and personalities most directly involved. Most importantly, the leadership in these organisations had to have confidence in the utility of the ODI team to be of assistance in these roles.

PM methods have served to inform both the ODI team and participating organisations. Specifically, ODI and its partner organisation in Udaipur, VBKVK, have been experimenting with the following forms of PM since 1994:

in *documentation*:

- the conduct of a number of village-level studies of GO–NGO collaboration;
- the publication of a number of Working Papers and Briefing Notes, both descriptive and analytical;
- the publication of a bi-annual review of 'Recent Developments in GO–NGO Collaboration'.

in *monitoring*:

- the establishment of a Forum at the KVK at which NGOs and GOs could learn of each others' activities and design collaborative action;
- extensive consultations with organisations and individuals.

We consider each in turn below, focusing primarily on those which have evolved in response to changes in the wider context of GO–NGO relations, in actors' capacities to engage with each other or in their perceptions of what individual PM techniques have to offer them.

Village-level studies (VLS)

The original concept behind VLS was that they should periodically record the principal elements of interaction among GOs, NGOs and farmers' groups, drawing out the main elements of success and failure and the reasons underlying these, and noting any evolution of working relations over a three year period. A number of villages were selected as case studies. The farmers' groups in these villages were primarily those formed with the support of NGOs, were predominantly of tribal composition (and so fairly homogenous) and were at various points along a trajectory generally comprising awareness creation, group savings and credit, input acquisition, processing and joint action in natural resources management.

139

The case studies utilise a participatory and iterative approach involving sequences of interviews, analysis of data, drafting of reports, reviews of results by participants to obtain feedback, and revisions. Rural communities, NGOs and GOs have been involved as reviewers and co-authors as well as informants. Specific components include:

- Assessments of past and on-going development efforts in agriculture, with specific attention to differences between GO, NGO and GO/NGO collaborative activities.
- Ongoing feedback on the progress of GO/NGO collaborative efforts, including performance of promotional activities, and attitudes, perceptions of participants.
- Analysis of events and developments in rural communities with reference to interactions with external agencies, GOs and NGOs.
- Detailed personal interviews and group meetings in rural communities, NGOs and GOs.

The VLS were also to monitor any changes in the quality of services provided, and help to determine whether farmers' groups in tribal areas can be an effective mechanism for making demands on government services. Baseline studies have now been conducted in three villages and 'repeat' studies in two of these. Emphasis has been placed on understanding situations and developments from the perspectives of the people involved.

Experience suggests that the original concept was not necessarily the most appropriate: it was impossible to predict in advance where the more illuminating changes in working relations which could be recorded within a 3-year period would take place. In the event, both the initial and follow-up VLSs have generated important insights, as have one-off observations in other villages. Two results of the VLS are: (i) they have illustrated the manner and enabling conditions in which farmers' groups can emerge; and (ii) they make clear the delays and complexities of dealing with a multiplicity of GOs and obtaining the support that is in principle available under government schemes for, among other things, agricultural inputs and the construction of water storage and transmission facilities.

The outcomes of the VLS include: (i) NGOs have engaged in intensive reflection on their activities; (ii) they have had their attention drawn to the collaborative dimension of their work; and (iii) reports have been prepared documenting their activities which otherwise it would not have been possible or likely for them to do.

At least one NGO, Seva Mandir, is on its own using a simple and direct form of process documentation as a means of recording agreements with government agencies and holding them to account. During meetings between Seva Mandir and a GO, the mere production of a notepad and pen was sufficient to intensify the GO's interest in the topics of discussion. Seva

Mandir is also using village studies as a means of internal learning. It is using a combination of documented and both formal and informal oral reports to chart areas of agreement with village groups and of agreement and disagreement among project staff in the hope of resolving areas of conflict over time.

Working Papers

Working Papers serve as a means of sharing information, analysis and ideas among participating organisations. The VLS were published as Working Papers, as were the first two numbers of *Recent Developments*. Others include a statement on PM methodology (Gilbert *et al.* 1995), a study of GO/NGO interactions in areas adjoining the Ranthambore National Park in Sawai Madhopur District (Lal 1996); and a paper on on-farm research (Gilbert and Sharda 1996).

The Working Paper series initially contained all the major written outputs from the PM activities. In 1995, it was decided to have *Recent Developments* as a separate series given its focus on reporting current events and the involvement of a partner organisation, VBKVK, in its production and distribution. Currently, Working Papers are devoted to special topics, such as on-farm research, and may include historical perspectives as well as discussion and analysis of concepts and experience. Shorter versions of Working Papers may be put out as special issues of *Recent Developments*. The Working Papers may not employ PM methods, but are a direct outgrowth of PM activities and utilise information generated through PM. Ideas for topics and potential collaborators for Working Papers grow out of suggestions by ODI team members, VBKVK and other participating organisations and individuals for more in-depth explorations of specific issues.

Recent Developments

This form of documentation has been associated with the most radical changes in GO–NGO relations in Udaipur. The initial aim was to be comprehensive: ODI staff documented any events in the district (and beyond, in the case of particularly illuminating examples) involving inter-agency collaboration in research or extension in agriculture or natural resource management. *Recent Developments* began as a means of keeping FF and a relatively small group of senior organisational representatives in Udaipur and Jaipur informed about events. The distribution list grew and *Recent Developments* evolved into a regular newsletter.

Changes in presentational style have been partly responsible for the increase in interest surrounding this medium since its first publication in mid-1994. Initially, it was produced by ODI with some support from VBKVK as a simple A4 typescript. In 1995, the effort became a joint undertaking with VBKVK. Beginning with the fourth issue covering the second half of 1995, the newsletter has been produced in a twin-column 8-page format. VBKVK also took responsibility for producing a Hindi version and for expanding both English and Hindi-language mailing lists. But there have also been changes in procedure and criteria for inclusion: there is now much more emphasis on having NGOs and GOs submit summaries of collaborative activities in which they have been involved. The newsletter has served to raise and discuss issues of interest to those concerned with collaboration in Udaipur District and beyond. VBKVK aims ultimately to become a compiler and publisher of the information assembled, though it currently still has to do considerable work in order to obtain appropriate material.

An example of the power of this medium became evident to all concerned in the third issue. We digress slightly to explain the background to this incident.

The component of the World Bank-supported Agricultural Development Project in Rajasthan entitled 'Privatisation of Agricultural Extension' had, since its introduction in 1993, been greeted with little enthusiasm by NGOs. The GoR Department of Agriculture (DoA) invited proposals in late 1994 from voluntary organisations to assume responsibility for extension activities in a specific area, including entire blocks or portions of blocks. As of early 1995, DoA had not received any responses to the invitation to turn over responsibilities for whole areas or parts of areas to NGOs. However, shortly thereafter a proposal from Morarka Trust was received and approved for Jhunjhunu District. The handing over of agricultural extension services in Navalgarh block in Jhunjhunu to Morarka commenced in May 1995.

The decision to contract agricultural extension services to NGOs caused considerable disquiet among DoA field staff. The Morarka Trust contract resulted in strong protests by a faction of the Agricultural Supervisors' (i.e. village-level extension workers') Union who expressed fear of redundancies and threatened to strike. DoA leadership reiterated that the decision to hand over the whole or part of the extension services in particular blocks to NGOs is in line with current GoR policies to economise on government operations and particularly to reach remote areas where there have been chronic difficulties in filling positions.

Partly in response to the protest from the AS's Union, government officials decided to limit the number of NGO contracts to one per district. However, this decision was initially not made public and a number of applications from Udaipur District in particular were submitted. Two of these proposals, notably from Prayatna Samiti (PS) and Ubeshwar Vikas Mandal (UVM) were prepared in consultation with the Deputy Director of Agriculture (Extension) in Udaipur and built upon the earlier contacts between these NGOs and the Department. The proposals specifically incorporated plans for the identification, training and deployment of para-extension workers to be drawn from the ranks of farmers in the villages in which these NGOs were already operating. The para-extension worker idea had been put forward by the Deputy Director following his participation in an international workshop on farmer-led extension organised by ODI and others in July 1995 (see below).

However, the PS and UVM proposals need not have been interpreted as a threat to public sector staff. The proposals specifically envisaged a continuing role for government extension staff to provide technical training for para-extension workers. In essence, UVM and PS were proposing a partnership with DoA, not a contractual arrangement under which DoA staff would effectively be withdrawn.

In January 1996, representatives of a number of NGOs from Udaipur and other districts were summoned to a meeting with government officials in Jaipur to discuss the proposals received. The meeting proved to be a waste of time, effort and money for the NGOs: government officials were poorly prepared for the meeting. They had not read the NGO proposals in detail and did not have the files available. The representative of one of the smaller NGOs from Udaipur which had responded to the invitation to Jaipur wrote a strong letter to the VBKVK as convener of the NGO/GO Forum on collaboration, complaining of their treatment at the hands of line department staff. A copy of the letter was published in *Recent Developments* in early 1996.

The letter was seen by senior officials in Jaipur, including the Secretary for Agriculture who decided to review the guideline of one proposal per district as a consequence. Subsequently, official approval was given for two proposals (PS and UVM) from Udaipur District. The announcements came just days before the conclusion of the financial year in March. PS and UVM were asked to come to Jaipur to sign the contracts. They expressed serious reservations since the contracts required matching financial contributions from the NGOs and specified that they should adhere to the 'Morarka Model' in the implementation of their programmes.

Despite these reservations, they agreed to sign the contracts. However, in April, the Additional Director of Agriculture (Extension) came from Jaipur to discuss the contracts with UVM and PS, making it clear that rigid adherence to the Morarka model would be required. The idea of para-extension workers was dismissed as unrealistic. It was clear from these discussions that the proposals from UVM and PS had not been carefully reviewed and had been approved largely because of pressures to act on various proposals before the end of the fiscal year. The impression given was that although the DoA had been given responsibility for the administration of this component of the ADP, including the processing of proposals, the procedures and capacity required for performance of these tasks were still evolving. The DoA had undergone three changes in the directorship since mid-1994 which no doubt contributed to uncertainties and a lack of continuity of effort and understanding of the programme during its early stages.

PS and UVM both decided to withdraw from the programme. In the wake of this withdrawal, PS wrote to the Secretary of Agriculture explaining their reasons for this action. In response, the Secretary wrote to PS indicating his appreciation of their frustrations and expressing the hope that they would search with him for a mutually agreeable way in which they could work together. This was followed by letters from senior line department officials indicating that the proposals were being considered on their own terms. These had clearly been written at the insistence of the Secretary for Agriculture, and subsequently led to the reinstatement of the contracts for PS and UVM with no reference to the 'Morarka Model'.

In the previous section, we noted the limited capacity and inclination among the actors – but especially in GOs – to draw lessons from unsuccessful development efforts. This issue of *Recent Developments* represented a landmark in the growing view on all sides that to do so was becoming legitimate, and that *Recent Developments* represented a powerful vehicle not only for documenting events, but also for hitherto neglected actors to express their views on them. Four sets of factors – not all of them immediately related to events in Udaipur – contributed to these changes.

- Most importantly, the fact that *Recent Developments* was an appropriate form of documentation for senior government officials at the highest level in GoR – i.e. the Secretary for Agriculture – and that he had responded sympathetically to concerns (especially NGOs' concerns) in Udaipur, and was apparently prepared to press mid- and lower-line department officials towards more flexible working arrangements with NGOs. The strategy of documenting events appeared to be working, and organisations in Udaipur, but also, more importantly, in GoR, had begun to use *Recent Developments* as a source of information and a means of expressing ideas. In addition, many – perhaps especially VBKVK and government officials in Udaipur – were pleased to receive positive attention from senior officials in GoR. This helped to legitimise their role in collaborative processes among organisations in and beyond Udaipur District.

- Further favourable attention was drawn to Udaipur by two other unfolding events: first, the involvement of an ODI Associate, who had been heavily engaged in PM, in a separate FF-supported study of GO/NGO relations in Rajasthan with a view to identifying alternative institutional mechanisms for promoting collaboration; and second, the establishment of a Working Group for the preparation of Rajasthan's IX 5-year plan proposals for improving NGO and rural community participation in agricultural extension and development. The Secretary of Agriculture invited three individuals from Udaipur who were familiar with and involved in collaborative activities in the district (including the Deputy Director of Agriculture and the ODI research associate) to participate in this Working Group. The resulting paper carried many of the ideas on farmer-led extension, on farmers as para-extension workers, and on multi-agency partnerships of the kind that were being promoted in Udaipur. It was warmly welcomed by the Secretary of Agriculture.

* The growing interest beyond GoR in the experimentation taking place in Udaipur with multi-agency partnerships for agricultural extension. Support from the FF had allowed three senior officials from the central government (GoI) to join five from Udaipur and one from GoR at an international workshop on farmer-led extension organised jointly by ODI, the International Institute for Rural Reconstruction and World Neighbors and held in the Philippines in July 1995 (Scarborough *et al.* 1997). The three GoI officials were the Joint Secretary for Agriculture, his predecessor, and the Deputy Director General (Extension) of the Indian Council for Agricultural Research. These were subsequently placed on the mailing list for *Recent Developments*, and their favourable comments have undoubtedly consolidated the VBS KVK staff's sense of ownership of *Recent Developments*, and enhanced their status in the eyes of peers.

* Although the amounts of donor funding provided for *Recent Developments* or for other types of documentation were small, there can be no doubt that the presence of FF and ODI staff in Udaipur helped local actors to gain access to senior GoR officials, and helped to attract the attention of officials from GoR, GoI and donor agencies to events in Udaipur.

The KVK Forum

In March 1994, at the request of one of the ODI Associates, a meeting was convened by the RAU Director of Extension Education in Udaipur at which NGOs and GOs discussed possible approaches to PM. Most, especially the NGOs, came away with more positive views than previously on the prospects of working together, and it was decided to hold further meetings. By mutual consent, these were held at the VBKVK and chaired by its Chief Training Organiser (CTO). These evolved into what came to be known as the KVK Forum.

Prompted by the more favourable climate for collaboration in the early part of 1994, a number of NGOs sought to establish collaborative activities with the extension services. These led to increased access to GO schemes by NGOs. Other joint activities set up as a consequence of the KVK Forum meetings included substantive input from NGOs into the design of training activities at VBKVK, and workshops on participatory planning.

In many ways, the Forum has fulfilled one initial purpose, namely of allowing the actors to become more familiar with each other (it had, incidentally, the unanticipated benefit of allowing the senior staff of GOs to become more familiar with each others' activities, since the only other district-level forum in which they met regularly was largely concerned with administrative matters). It continues to meet a second purpose, i.e. that of providing a semi-formal setting in which NGOs and GOs can monitor and express their views on recent events – at times by reference to the types of process documentation introduced. From time to time it has served other purposes: it has, for instance, allowed discussion of the purposes of the Agricultural Research Fund provided by the FF, and the modalities of screening applications. Many participants feel that, strictly as a venue for discussion and exchange, two years after its establishment the Forum is in need of a change. Some feel that this might take the form of occasional presentations by 'outsiders' on matters of GoR, GoI or donor agricultural policy. However, such events take time and effort to organise, and, as we discuss below, those who would be ultimately responsible for their organisation can obtain stronger recognition for less effort in other PM activities. Apart from facilitating specific joint activities, the broad area in which the Forum has made progress is in providing NGOs, especially the smaller ones which tend to be overlooked in district-level discussions and do not have the resources to visit the state capital frequently, with a platform on which to voice their concerns. Added to this, the Forum has from time to time requested its Chairman to write to other officials on its behalf. Thus, for instance, it asked him to write to the Vice-Chancellor of RAU expressing concern at the poor representation of the RAU at the Forum and, by implication, its apparent lack of interest in responding to some of the technical concerns raised by those present. In the broader context, the joint setting of the agenda for Forum meetings, the detailed recording of minutes and the opportunity at subsequent meetings to check whether agreed follow-up had, in fact, taken place, are taken for granted in many settings. However, in the Indian context they are all examples of a growing confidence among those concerned with technical change appropriate to the needs of the rural poor that they *can* call on services that are theirs by right and can expect to receive positive responses.

Individual consultations

Individual meetings, both formal and informal, are easily the most time consuming and arguably the most important dimension of PM work. Although formal meetings such as an inter-organisational forum can be focal points of intense interaction, these tend to be episodic. A great deal of communication takes place outside of such meetings. PM 'services' can be extended to include a range of less formal communications among participants, including those that are private or semi-private in nature. Verbal communications as well as letters between specific individuals are very important elements.

PM work can quite naturally 'spill over' into the provision of a range of advisory services to individual organisations. Such services may include assistance in drafting letters, proposals for funding and documentation of activities as well as general brainstorming on strategy and tactics. These kinds of services are extremely important in facilitating collaboration, especially since some organisations may lack the capacity in certain situations to communicate effectively otherwise. This is especially true where a large GO or NGO is interacting with a small NGO or village group. In such situations, there may be a need for an intermediary, such as can be provided through PM, to better ensure that messages are delivered and understood.

Prospects for PM

This chapter has argued that PM initially faced a difficult situation given:

- the limited tradition (especially in government) of drawing lessons from success or failure in the *processes* of development;
- the fact that PM is only taken up by NGOs or GOs to the extent that it generates benefits for them. PM was not linked in any way with potential financial benefits from the agency (FF) supporting it. Any such benefits would therefore have to be generated in the context of NGOs' and GOs' routine working relations, or from special funds made available by GoR, GoI or other donors;
- the strong competing demands on the time of NGO and GO staff. PM would have to generate substantial benefits, and quickly.

The previous section demonstrated that interest in the different modes of PM that had been introduced had varied between NGOs and GOs, that the interest in them depended in part on their contribution to the resolution of immediate difficulties, but also very significantly on their wider contribution to the status of the organisations and individuals involved, and that such changes in status are intimately linked not only with PM but also with other changes beyond PM.

Two questions remain:

- first, what are the prospects for sustained deployment and evolution of PM in the setting discussed, and for its introduction elsewhere?
- second, what is the appropriate role of external agencies involved in PM?

Taking each in turn:

Is PM sustainable?

Given that PM has been in operation in Udaipur for less than three years, some observers would argue that the answer to this first question is premature. Yet certain trends are already clearly discernible:

- the successful application of PM depends to a large degree on positive response to PM from senior government officials, and on their ability to ensure that mid- and junior-level line Department staff implement agreed policies;
- PM requires time, effort and patience where demands on key actors' time are already high. Encouragement from senior officials is important, but cannot substitute for work at the coalface;
- it deals with sensitive issues, and so the risk of causing offence to influential individuals is high.

As we argued in the previous section, 'success' depends at least as much on positive change in the wider context as on PM methods themselves. In this context, the portfolio of PM techniques is evolving rapidly in Udaipur. Those which have fulfilled part of their initial function (e.g. the Forum) are likely to decline unless new roles can be found for them; those yielding results of most interest to certain types of organisation (as with village studies), may be replicated by them and, as we saw with *Recent Developments*, those showing promise are adapted and expanded.

PM has helped to stimulate more flexible approaches in line Departments to NGOs, bringing positive results in the context of specific joint actions. However, its underlying impact (though intimately intertwined with other changes) has been on the levels of confidence of organisations – both GO and NGO – operating within the district. In terms of their relations with each other and with the State capital, this may express itself in further evolution of PM methods (such as *Recent Developments*) which have already proved themselves. Plans are already in hand to produce 'special numbers' on selected themes, and the regular issue appears set to take on more of the appearance of a newsletter – even newspaper – as local organisations feed information into it more readily. But enhanced confidence has also manifested itself in fuller use of the conventional armoury of democratic pluralism: ability to negotiate agreements; setting of agenda for meetings, the taking of minutes and the assignment of responsibilities for follow-up, and so on. ODI team members have provided informal advisory assistance to a number of NGOs in such areas as communications, networking and the preparation of proposals for funding. In this context, such assistance serves a nurturing function. Overall, there has been a transition from the initial concept of a tight cycle of action, documentation and reflection to a more flexible set of arrangements in which some elements of this cycle remain, but overall a larger number of agencies are involved and these monitor events more informally than originally conceived.

PM can assist in measuring the impacts of collaboration on organisational performance and on rural communities. However, PM's priority concern, as well as its area of comparative advantage, is in dealing with the present, and specifically in assisting participating organisations to address the immediate problems and challenges posed by working together collaboratively. At the same time PM activities generate considerable amounts of information which can be selectively used for a range of purposes. PM does not offer any magical shortcuts in such areas as impact assessments, and existing methodologies are perhaps best employed for such topics.

As far as the spread of PM methods to other areas is concerned, evidence from Udaipur suggests that an external facilitating agency has a potentially important role to play, not least in gaining the attention of senior officials and that a Forum for face-to-face meetings is important in order to reduce barriers and misperceptions in the early stages. As confidence grows, a journal such as *Recent Developments* may become the most important single vehicle. However, PM is unlikely to spread easily for a number of reasons:

- there is little tradition of promoting free flow of information in the Indian context, especially, but by no means exclusively, in the public sector, where the tendency to 'privatise' information and so remove it from the public domain is a natural component of rent-seeking behaviour;
- it is not clear by what criteria the boundaries of groups of NGOs and GOs wishing to explore the potential of collaboration might best be defined. Some might argue, as in the case of Udaipur, that because it is the lowest level at which most government departments have budgetary allocations, the district is the most appropriate level. However, only in few districts will officials be found who have the necessary initiative and skills to respond positively to opportunities for collaboration. Training in interpersonal skills may help to some extent, as may a greater element of decentralised authority. However, the spread of PM is likely to remain patchy.

Collaboration is clearly the type of undertaking which can benefit from PM-type services to facilitate communication at several levels. Our experience strongly suggests that PM approaches must be prepared to change over time as the collaboration experience proceeds through various phases and as participants gain an increasing familiarity with one another. The phase of collaboration influences the nature of the PM services required and the role of a 'third party'.

The role of external organisations

We argued in the previous section that the uptake and impact of PM methods will depend on the extent to which senior government officials favour change and are able to press through its implementation by line departments.

FF and ODI have contributed to the climate of positive change by inviting senior government officials to the international workshop on farmer-led extension, and by networking, both formal – through the Agricultural Research and Extension Network – and informal through correspondence, routine meetings and so on. Local organisations and individuals are primarily responsible for motivating change, (in some cases) preventing it, and adjusting themselves to it. However, it is partly also through networking by outside organisations that change comes about, and vested interests, rent-seeking opportunities and the positions of organisations in relation to each other all redefine themselves.

ODI and FF also contributed to an exploration of the characteristics of a range of different PD and PM approaches by organising a workshop in the UK in April 1995 attended by some 50 persons from academic and other institutions in the UK interested in adapting and applying these approaches. The output from this workshop was the genesis of the present book on PM.

A key question concerns whether, how and how far an external organisa-tion should take the lead in analysing the findings of PD and PM. Where these techniques are applied in the context of donor-supported projects, there is undoubtedly a ready prospect of feeding the results of such analysis into the modification or new design of such projects. PM is being used by indigenous organisations as part of an effort to improve their individual performances as well as working relations among them. As such, the respon-sibility for analysis lies primarily with them. Some may wish to record the results of their analysis in writing; others may prefer to express them orally. In all events, interactions among and within indigenous organisations are likely to be conditioned by nuances of behaviour at times barely visible to outsiders. In such cases, it is all too easy for outsiders to misinterpret events or to draw conclusions insensitive to the positions of key actors. In these settings, outsiders are best advised to leave analysis to local organisations, restricting themselves to support and encouragement in the use or adapta-tion of PM techniques by those wishing to use them, and to helping to create a positive context for change.

Notes

1 The authors are all associated with the Overseas Development Institute (ODI), London and constitute the ODI team involved in PM activities in Rajasthan.
2 A more complete treatment of FF's role is the focus of Chapter 7 by Ruth Alsop in this volume.

References

Gilbert, E.H. and Sharda, H. (1996) 'Farmer participatory research for NGOs and Farmer Groups in southern Rajasthan: an assessment of the (lack of) progress to date and a proposal for an NGO partners programme', Draft Working Paper, BKVK, Badgoan, India.

Gilbert, E.H., Khandelwal, R. and Ballabh, P. (1995) 'Process monitoring methodology: preliminary concepts and approach', *Working Paper No. 1*, Udaipur: Vidya Bhawan Society Krishi Vigyan Kendra.

Lal, J. (1996) *BAIF Case Study*, Draft case study prepared for PRADAN.

Professional Assistance for Development Action (PRADAN) (1996) *Towards A Relationship of Significance: Interim Report on the Study of Relationships Between Government and NGOs in Rajasthan*, New Delhi, mimeo.

Scarborough, V., Killough, S., Johnson, D. and Farrington, J. (eds) (1997) *Farmer-led Extension: Concepts and Practices*, London: IT Publications.

Part 3

PROCESS MONITORING AND POLICY REFORM

9

THE RESOLUTION AND VALIDATION OF POLICY REFORM

Illustrations from Indian forestry and Russian land privatisation

Alan Rew and Angelika Brustinow

The policy process – an introduction

Policy making is about the complex and continuous process of adjusting the 'value system' to the 'reality system' and vice versa . . . when viewed in this light it becomes much more relevant to talk not of goals and objectives that are achieved, once and for all, but of norms and standards that are maintained or modified over time (policy 'problems' being resolved rather than solved).

(Gregory 1989: 227 commenting on the work of
Sir Geoffrey Vickers)

The combination, in this chapter, of insights gained from Russian land privatisation and Indian forestry reform starts from a faith in a particular *theory* about the policy process and the role of monitoring. Our aim in monitoring processes of policy change in Russia and India has not been to provide a simple record of policy making or a piece of social science interpretation for the academic literature. Two separate case studies would have done the job of record-keeping more efficiently; and most social science commentaries concentrate on the substantive content of policy measures rather than the procedural issues of reform which concern us here. We do not concern ourselves so much here with the aims of policy but with the process of monitoring and investigation which follows the decision to intervene and to provide technical monitoring assistance to a policy reform intention. The monitoring arrangements we report here highlight for us the urgent need to view policy as a process of incremental mutual adjustment, as the outcome of multiple stakeholdings, and as the embodiment of explicit and implicit norms and standards about sensible courses of action.

In order to understand the process, and our involvements, we have drawn on both practical policy monitoring experience and the 'policy process' literature of 'incrementalism' (Lindblom 1980) and 'discursive institutions' (Sabel 1994). Our own aim is explicitly practical rather than theoretical although the positioning of the practical subject within theoretical terms is critical in the initial stages. Where we do elucidate theory it is less to elaborate the concepts of the policy process literature than to describe the discursive hiatus between the process monitor's role and the 'ends–means' views which dominate in the practical world of development aid assisted policy reform. In the dominant discourse, monitoring is usually seen as the deployment, within a system, of economic and sociological tools to calculate and evaluate outcomes and effects. In our view 'process monitoring and research' is best seen as embedded within, not as something beyond or outside, the institutions which guide the policy reform process; it is part of an institutional and policy learning process rather than a set of tools for project assessment.

The growing institutional complexity of aid allocation into policy reform and programme packages has paralleled experiments in the 'process' planning of aid projects. The changes in aid modality have arisen largely from failures in the field, and the lessons drawn from them, and to a limited extent because of developments in the academic analysis of public administration. Emphasis has shifted from the tightly controlled 'blueprint' project with its 'once and for all' enhancements of physical capital and economic efficiency to 'experimental' or adaptive project plans which focus on institutional change, flexible budgets and the achievement of project milestones and contingent indicators of effect and impact. Measurement of impact is especially important in order to judge the quality and direction of further aid inputs. The experiments have been stimulated by many experiences of project implementation (Rew 1997). There are references to the academic literature (for example, Rondinelli 1983) in the documentation of 'process planning' by aid donor agencies and their staff. But the major emphasis in the donors' procedural statements has been on the pragmatic need to innovate in the management of the project cycle, create flexibility and move the agenda from physical capital development to the growth of human and institutional capital. Eyben and Ladbury (1995) and Coles (this volume) have discussed the identification, appraisal and implementation of 'process projects' and the aims of the donor (the UK's Overseas Development Administration, now the Department for International Development) in supporting 'process' projects. By and large these statements aim at the pragmatic justification of flexibility and responsiveness in project cycle management in order to meet primary stakeholder needs; the nature of the policy process and the conceptual and methodological issues it raises are left largely unexamined.

Another untheorised development concerns recent changes in monitoring and evaluation methods. The shifting focus of international aid – from large infrastructure projects to policy/sectoral reform and adjustment and programme aid – ought, in turn, to lead to re-examination of the efficiencies of monitoring and evaluation methods. As the aims of aid have become more complex and conditional upon government reorientation and response there is a need for more complex and politically sensitive monitoring. There are often major conundrums and ironies which need sensitive understanding. For example, despite their apparent agreement to policy reform in order to secure aid packages, recipient countries would usually much prefer to have investments which promise employment and other visible and physical impacts. The monitoring system must try to accommodate the need to spot both recognised impacts and those impacts which, although not crucial to the aid agreement, are crucial to internal political stability. At the same time, to ensure political sustainability the monitoring system must be able to evaluate outcomes from the shifting perspectives of a range of project *opponents* as well as potential beneficiaries. The procedural answers may be to: include explicitly *political* variables in the monitoring or employ political scientists as part of the monitoring task force; initiate retreats and workshops with a variety of participants; and build flexibility into the project design in order to respond to the rapidly changing political agenda. The theoretical implications of monitoring from the standpoints of a multiplicity of policy stakeholders, or the implications for the monitors themselves, have not been addressed. Many of the theoretical issues centre on the balance between the resolution of piecemeal policy problems as they arise through action research for multiple stakeholders and the establishment and protection of an independent account of policy impact.

Practical project and policy decision-takers rarely have time to theorise. Even so, there is surprisingly little consideration given in the process project documentation to the nature of the monitoring process and the monitoring roles involved. Since the aim of process monitoring is to monitor social relations there ought to be more than minimal reflection on the nature of the social relations around the monitoring process; we need a theory of the monitoring process.

To understand the roles and system changes involved it is worth starting from a review of the supposedly fundamental distinction between 'Blueprint' and 'Process' projects. Now both of these terms are odd; in the first case because the term is intended to be somewhat derogatory and in the second case because the term has been shortened – from *Learning Process* (Korten 1980) to simply *Process* – and changes its referents and meaning as a result.

Blueprint may imply a rigid and hypothetical, desk-bound engineering of design and outcomes. It is, in that sense, incomplete and therefore somewhat dismissive of engineering skills. It implies an engineering which lacks project and contract management subsequent to design, whereas conventional engineering would necessarily include all aspects of both design and implementation. Additionally or alternatively, the 'blueprint' idea implies possible stupidity – the thought that design and intention, provided these are detailed and precise enough, can control implementation and outcomes and therefore the idea of a 'blueprint' alternative is used as a strawman to create the opportunity for 'process' projects.

Learning process projects usually gets shortened to simply *Process projects*. This probably involves a major change in the referents. 'Process' is a continuous or regular series of operations carried on in a definite manner. Thus it, too, has engineering connotations – e.g. the Bessemer process – or legal ones – i.e. to institute a process of action against. Thus a desirable end state, known at the outset of the project, is already implied. Many, indeed, would say that a *Vision* of the project end and the 'definite manner' of the operations is needed before there can be a process project at all. As Rick Davies writes (this volume), this type of process project is deductive – there is a prior conception of a desirable end stage – the means can be varied and rescheduled to deal with uncertainties – but these means are deduced from the end state desired by a vision or a predetermined plan. Davies contrasts the kind of project that depends on a prior conception of the desirable with what he calls evolutionary, open-ended projects. His documentation of the 'process monitoring' system adopted by CCDB in Bangladesh describes a system and corresponding monitoring framework that accepts and expresses diversity and uses internal competition, local salience and situationally selected critical events to explore the boundaries of organisational change and social outcomes. This is the essence of a learning process organisation in which 'diversity becomes an opportunity rather than a conundrum' (Davies, this volume).

The problem with human diversity is that the social groups, categories and actors which constitute the diversity usually express or develop differing interests and ascribe contrasted meanings to events. This is why diversity sets a conundrum for economic and social development. Diversity implies economic exchange and markets and the adaptation of knowledge and skills to local conditions and changing circumstances – in short to change and innovation in economic relations. Yet these changes and innovations create uncertainties about benefit – and can lead to the breakdown of the underlying exchanges if worries about loss of economic niches and livelihoods predominate. Sabel (1994: 231) also argues the need to reconcile economic growth through learning with the determination, by the transacting parties, that the gains from learning are distributed as agreed. Sabel regards this as *the* central dilemma of growth. The Japanese system of industrial organisation resolved the problem of reconciling economic learning and the monitoring of gains

> by creating institutions that make discussion of what to do inextricable from discussion of what is being done and the discussion of standards for apportioning gains and losses inextricable from apportionment . . . these institutions transform transactions into discussions, for discussion is precisely the process by which parties come to interpret themselves and their relation to each other by elaborating a common understanding of the world.
>
> (Sabel 1994: 231)

The same common understanding is needed in the public bureaucracies of national and international development. There, too, innovations threaten the principals' control of their subordinate agents. Procedures for the implementation of projects become so ambiguous that agents develop their own independent aims and ends, not those of the supervisors or senior management who approved the design, and implement the procedures in a way that is difficult to detect and sanction. The need to understand these agency processes 'from within' has been stated by Chambers (1994) and attempted, for example, by Apthorpe (1986). Escobar (1995) questions, for anthropologists, whether an independent, critical perspective can be developed through the participant observation study of aid industry organisations; he sees the attempt as essentially compromised and compromising. Escobar's case is suggestive of the way that the underexamined assumptions of power and a dominant discourse can undermine critical investigation, but the incorporation and compromising of the investigator is not inevitable. A study 'from within' could be highly subversive of the development organisation's purposes and methods. Moreover to assume the compromising and incorporation of the investigator of internal process from the very beginning avoids the need to recognise the practical purposes and implications for aid delivery systems of studying those social relations which encourage or prevent innovation within an agency. The relative lack of first-hand studies on the organisation of the development aid policy process need not constrain all progress on the development of relevant theory. Reference can be made to, and help found in, the extensive North American and European literatures on the nature of the policy process (for example Hill 1993). These literatures include many case studies and theoretical discussion of the nature of the policy process which can be useful to development policy analysis and its institutional underpinning. They confirm that while policy analysis is often uncomfortable – in Wildavsky's (1987) terms it means 'speaking truth to power' – it is nonetheless feasible. The unremitting, radical worries about the dangers to ideological purity which analyses of the Escobar kind generate may be misplaced; there is room to manoeuvre to shape the policy

process and policy content while acting in a facilitating and process monitoring role.

The available general theory and first-hand experience in development
policy indicate that attempts to intervene and support initiatives in policy
reform must build in mechanisms and resources to monitor achievements
and to validate them. The outcomes of policy reform are usually not easy to
discern – they need careful sifting – and once discerned need to be recognised
and confirmed. This raises the need to see in the monitoring process an
element of *validation*, a mechanism that creates agreement and consent
about results which can often be contested in their detailed composition.
And if *validation* is required, then what models of the validation process are
most appropriate? The mechanism that comes readily to hand to the university academic is the meeting of all examiners in an examination board with
an external examiner (read process monitor?) present and guiding the
discussion and outcome. Another mechanism used in engineering and the
management of production and service is for a semi-independent member of
the system to sign off that the output has been vetted by a Quality Assurance
procedure. In development project planning the usual procedure is the periodic independent and/or joint review mission. As projects become more
process oriented, however, there is need to consider the role played by the
project monitor in steering the processing of emerging project results
towards agreement and towards their validation. This is a crucial role
because the action of resolving standards and maintaining norms for project
discussion is critical to the agreement of project and policy objectives. As
Lindblom (1959: 84) argues: 'policy objectives have no ultimate validity
other than that they are agreed upon'.

The next part of the chapter puts some of these ideas about resolution
and validation, rather than purpose and achievement, to the test. Cases of
learning from policy reform in two big marker countries – Russia and India
– are examined. These countries are important in geopolitical terms and
because of the levels of development aid involved. Moreover, liberalisation,
privatisation and major change in the standards and processes which govern
social and public institutional life are key current topics and areas for policy
experimentation.

Learning from policy experiments in reform

Perhaps the first question to ask is – why monitor at all? What is it that is to
be assessed; and why monitor the process and the means of intervention
rather than wait and simply evaluate the outcomes of the intervention at the
end of the process? One important and conventional answer relates to the
sustainability of the investment or policy measure and its contribution to
well-being, human development and relief from poverty. The investment or
measure must have a prospect of continuing to achieve its aims beyond the

immediate period within which it was introduced. It will not be sustainable if its aims are not being achieved because of unexpected and unwanted constraints or if secondary consequences undermine any of the bases on which the investment or measure was initially mounted. These bases may involve the sustainability of the natural resource, the supporting social capital and trust or the financial viability of existing exchanges and production. Forestry-based livelihoods are one such area of conflict of geopolitical salience and India a key site; the drive towards privatisation at farm level in Russia also raises deep questions of sustainability since we consider continuing questions of regulation and also of individual household incentives to produce in the face of declining standards of living.

Another answer concerns the aid donor's accountability to taxpayers in its own country. Political realities mean that as public finances come under increasing pressure, interest groups lobby for and against overseas aid and the donor agencies must have evidence of their impact and effectiveness. There is another reason for assessing policy reform measures in a British funded aid project. Is privatisation a lasting issue for transitional economies and major Asian ones – or of limited interest, mainly to Britain and to a Thatcherite Britain of the 1980s?

In Russia there is widespread suspicion of, and resistance towards, policy reform in the country. The suspicions include those which arise from what are seen as alternative social science or ideological explanations imported from the rest of Europe and the United States. There are also the concrete consequences of reform to fuel the fears. In a country which has always stressed the merits of economic autarchy and independence from the capitalist West, the population now finds that at least 40 per cent of foodstuffs are imported. People are worried that privatisation and the restructuring of large collective farms will make the country more dependent on imports. The West was also the historical enemy of the Soviet regime. People worry that listening to Western advice on land privatisation will distort Russian needs and judgements and lessen the country's ability to cope with its own poverty or to fulfil its destiny.

Another conundrum in monitoring in general, but especially in process monitoring, lies in the conflict between general intentions and aims and short-term political risks and realities. The major stakeholders (such as the executive agencies) and the clients (the beneficiary population) all want immediate and visible results to justify the programme and spending and its social, political and economic risks. The donors also want immediate results but they are also concerned with long-term change and impacts and the continuing and generalised process of reform. The conundrum is always about pace – does the project risk sacrificing the long term through visible short-term concessions required to sustain all participants' interest and motivation? If there is no other reason for it, this critical need to balance the short and long terms at all stages of the intended changes is a key reason for the continued monitoring of the process of reform. Monitoring cannot be left to the 'end' when reform is complete because agreements and impacts are always contextual and at least some institutional history is necessary to understand the movement which has taken place. Nor is it wise to divorce the collection and analysis of data to meet the aims of a reform project from its release during the reform process. Decision-makers need guides and justifications for the risks and possibilities of benefit they sanction at all stages of the project, not only at its end.

These monitoring requirements are seen in heightened form in the Russian case study reported here. In order to raise what was originally a pilot project into a national policy reform initiative, there needed to be a series of analyses and justifications to support the case. Results from the process monitoring were needed to demonstrate the potential benefits which would accrue and the appropriateness – in social, political and economic terms – of the land administration and titling arrangements being used in the pilot.

This evolution in Russia has similarities with the case of Indian forestry reform, but there are also important contrasts. The Russian Prime Minister's endorsement of the Nizhiny Novgorod model of land reform as 'the' Russian model had only a nebulous, and 'non-obligatory' status until as late as March 1996 when President Yeltsin issued Decree 337 establishing deadlines for the issuing of land shares and contracts. There are now a number of Russian models based on differing local circumstances. Piecemeal policy experiments in Indian forestry were first conducted in one or two states (notably in West Bengal), generalised by means of a new national forest policy to the nation as a whole and then implemented in particular states through aid-assisted development projects from circa 1990 onwards. Some states were early pioneers and could propose models for national adoption. The declaration of a national policy which was independent of the experiments and evolved systems at state level, however, gave an importance and legitimacy to the subject which was a valuable resource for hard pressed reformers in each state. The national policy validated the state experiments and pioneering programmes. It also reinforced a tendency to resolve policy controversy by a 'stroke of the pen'. In at least one state, implementing legislation and regulations have been permissive and non-prescriptive and have allowed further reform processes and innovations to unfold. In another state, as we shall see, the rather premature adoption of excessively detailed regulations helped freeze the reform process and turn 'process' into administration. 'Process' implies the evolution of solutions and resolutions which are then known to be provisional for a period, and subject to scrutiny and amendment. The proclamation of an edict or set of regulations signals that, for a long while at least, the search for solutions and innovations is over – government has decided. If this occurs too soon it can abort the process of reform.

The task of the reform and monitoring process is to find the specifics of implementation through which distinctive local features and the necessary local dynamics can be generalised and further developed into guidelines, models and procedures. The aim is to identify both generalisable beneficial outcomes and local variants which can create a sense of local pride and ownership in the programme activities. So in the Russian land privatisation case we have, in addition to the Nizhniy Novgorod model, the 'Oryol' model, the 'Reform experience of the Don region' and others. One reason is that some national advocates of land reform wish to distance the new ventures from the image of ultra-radicalism and Western pressure associated with the Nizhniy Novgorod model and the criticism that the Nizhniy Novgorod model has been forced on the rest of Russia. Another aim is to stress local characteristics to increase the chances of local acceptance. In the case of Joint Forest Management ('JFM') in India, although the initial experiments were located in specific states – for example West Bengal – and these states now have a special reputation and distinctive capabilities; there is little sense of a rolling programme of institutional search and innovation. Each state now sets about implementing the JFM template according to its distinctive needs but without the sense of search and innovation which shines through the Russian examples.

So whom – given these various demands on the monitoring system – do we monitor for? There are clearly many interests and stakeholders which have to be serviced. This multiplicity of demands provides another reason for a monitoring system which is distinctly process-oriented. In other words, there should be an adaptable, open-ended monitoring system able to respond to the different, emerging and changing demands. Process monitoring becomes an essential element in the reform process, not a luxury for wealthy, high-profile projects.

Once it is decided that process monitoring is required the next step is to decide how to balance the various demands on it. How much of a possible trend can be revealed, and to whom? How does the monitor keep faith with all the stakeholders while revealing 'facts' and 'analysis' as the programme's hypotheses are answered? These questions raise serious and unavoidable issues about the reporting and publication of process monitoring results. They also cause process monitoring to confront its relations with the project's 'PR' or public relations machinery. To what extent should process monitors seek to influence the routine dissemination of official information about the reform process and its pilot projects?

The experience of Russian land privatisation

Background

Following the political events of the beginning of the 1990s, fundamental changes are taking place in the Russian economy. Russian agriculture, like other industries, is in a state of transition from a centrally planned and controlled system to one based on diversified property rights and free markets. The state (*sovkhoz*) and collective (*kolkhoz*) farms of the former Soviet Union had an average size of 8,000 hectares and were substantial and often diversified enterprises. Since 1990, changes in policy and law have opened the way for the restructuring of *sovkhoz* and *kolkhoz* into smaller farm enterprises owned and managed by entrepreneurs. In today's Russia, few issues are more certain to arouse emotion and provoke heated political debate than the issue of private ownership of land. This present account of the reform process and its monitoring draws heavily on the analysis by Brustinow and Turner (1996).

Since late 1993 the British Know How Fund has been working in Russia to design and develop a simple, fair and legally defensible generic model for Agricultural Land Privatisation.[1] The work has been implemented through a partnership with the International Finance Corporation (IFC), an arm of the World Bank. Using foreign privatisation advisers and Russian agrarian experts, work started on devising a 'bottom-up' choice-driven privatisation method which is accepted as workable and fair by the farm community.

The model creates new private farms and farm businesses by dividing the state and collectively owned land and property among all qualifying people, defined by Russian law as living present and former members of the farm collective. Specially created entitlement certificates give qualifying members of the farm the purchasing power to 'buy' land and property. Land entitlement certificates are equal in value – managers and milk-maids receive equal portions. The value of property entitlement certificates is calculated according to a person's tenure and salary history. Pensioners as a group play an important part in the reorganisation since approximately half of all collective farm residents will be retired workers and are entitled to land and property entitlements. It should be noted that the farms were also social institutions serving a community of usually several villages and providing its members with cradle to grave housing and social services, food, shopping and cultural amenities.

Privatisation under the model programme is achieved through a sequence of steps. Entitlement certificates are distributed; an information campaign is launched to educate shareholders on how their certificates may be traded, leased, bequeathed, etc.; how shareholders can 'buy' a portion of the land and property lots – based on existing operational subdivisions – using their certificates. No foreigners or other outsiders can buy land and property; no cash is involved.

As the project has evolved it has come to focus on the following technical and institutional aims:

- to prove that the approach to land privatisation which has been adopted – the 'Nizhniy Novgorod model' – can be successfully managed and implemented at a regional level using local resources;
- to use lessons from this pilot and regional work to develop quality Russian Federation capacity to train and sustain Russian managers and management systems in the approach;
- to refine the detailed mechanisms involved through continued process experimentation and monitoring work;
- to train specialists, officials, and farmers at the local level to establish the capability to reorganise in a number of regions;
- to maintain a political dialogue to encourage local-level Russian Federation ownership of, and support for, land privatisation.

The first step in the process to arrive at these broad aims was a pilot project in one region (Nizhniy Novgorod) which created and tested a detailed mechanism of reform and developed the supporting documentation and legislation. Components have been added to try to strengthen the process, such as advice and support for more appropriate methods of on-farm communication, agricultural and farm business advice for farms after privatisation and support for the *Oblast* (District) Department of Agriculture (DoA) for addressing policy constraints. The method and legislation of the Nizhniy Novgorod model was then endorsed by Prime Minister Chernomyrdin by Governmental decree in March 1994 as a desirable national programme.

After the first pilot stage, the KHF/IFC project inputs concentrated, from summer 1994 onwards, on the development of Russian management capacity to implement farm privatisation on a wider scale. Lessons learned from the first pilot about how to refine and improve the approach and, from the wider programme about how to institutionalise it, were incorporated in the new phase. Work also started in Southern Russia (Rostov-on-Don) to test a number of new applications: the use of the 'NN' model on much larger farms and in one other region; a lower-cost approach to management and implementation; and to show that an approach started in a zone of quite risky agriculture could be replicated in Russia's most productive agricultural regions.

A third phase started in mid-1995. The main purpose of this third phase of the project was to further refine both the detailed mechanism and the methods for building Russian management capacity and to validate them through *close monitoring*. Several hundred Russian officials have received training in how to implement the model programme. More than 300 farms have been privatised and reorganised under the model programme in more than 10 *oblasts* with KHF/IFC assistance.

The process of project development and policy reform has led to a project whose purposes are to reform inefficient Russian agriculture, ensure the viability of the food chain and promote a market-based rural economy through the privatisation of farm land and the restructuring of agricultural enterprises. Four activities are helping to ensure the success of this objective:

- an increase in the rate of privatisation of state and collective farms through the creation of fair, open, and replicable procedures and guide-lines;
- the development of federal and local legislation to bolster and simplify the process of land privatisation;
- checks on the quality of reorganisation and the level of awareness of the farm population;
- the facilitation of a system of post-privatisation support to ensure the viability of new farm units.

In the 1996 Russian Federation elections Land Reform issues were one of the most important elements of President Yeltsin's campaign and a postcard carrying his signature and containing a brief summary of shareholders' rights and options based on the project model was sent to every second entitlement holder in Russia. But the project and process continues to face very difficult opposition. Farm privatisation is one of the most sensitive and difficult issues in post-communist Russia. Progress of the project is followed very closely by the media and by the opposition parties at local and federal level.

Key monitoring issues

The project has always faced a very inquisitive and sometimes hostile press. In the early phases of the project the form of information gathering and analysis was strongly driven by the need to provide answers to the opposition to the project. For example, the project used communications specialists in order to run a PR campaign including preparing articles for the press, organising professional PR and the targeting of influence and information on key opponents.

The need to provide 'impact' information to reach a wide variety of different stakeholders has always been accepted but is still to be fully evolved and established. Indeed some stakeholders have decided that they can only trust their own impact assessment. For example, in the early stages the (conservative) Agrarian Committee of the Russian Parliament (Duma) decided to send four sociologists to monitor, over a year and a half, the impact of agricultural re-organisation in the region of Nizhniy Novgorod.

The importance of providing information to such influential, but sceptical, groups of stakeholders was a major factor behind the initial design and approach to process and impact monitoring. In the early stages of the project there was a tendency to target the 'outsiders' with information that was sometimes very selective and overly positive. Presentation was usually excellent but the underlying base of information and analysis tended to be poor. For example, the economic information on yield on the pilot farms was not very good and the comparison with official information about non-project farms used data that most people knew to be unreliable.

The monitoring approach has to be able to produce information speedily and flexibly to meet different demands. This requirement encourages the use of case studies which are very useful for targeting the media and exciting the interest of farm people. Much impact monitoring has been on a sort of 'campaigns' basis (for instance because of the pressing need to get information for an important Russia conference, or for the donor's processing and decision-making requirements).

Forecasting the set of impact variables that different groups are interested in is very difficult. The set of variables favoured by the initial donor stakeholders (KHF, IFC, etc.) were not necessarily those that were most influential in affecting Russian opinion about the process and in stimulating take-up. Important impact variables ranged through a wide list of issues about the privatisation process itself (legality, fairness, transparency) to the impact of privatisation on the social services (access, quality of service) to issues about economic impact (profit, yield, financial stability, cost of transition) to issues about the cost of implementing the process itself (quality of privatisation versus cost of foreign advisers).

The KHF was, in the first instance, most concerned about the privatisation and reform process, assuring its quality and fairness, and the ability of the farm population to make well-informed choices when exercising their ownership rights. KHF was probably less interested in yield and economic impact but there was still a strong demand for this information from other parts of ODA (now DFID) and from the Russian stakeholders.

Those economists and donor representatives most concerned about accountability questions and wanting to view likely 'end-of-project' results were disconcerted by the project's difficulty in isolating a set of 'hypotheses' on which to base impact assessments. But these 'difficulties' arose partly from the frequently changing demand for information about different impacts from different groups and also because of the demand from the project insiders (including the KHF) for a very flexible process which would allow new and difficult issues to be exposed. The wish to run 'PR' campaigns and to include a multiplicity of stakeholders, including declared opponents, within the project's scope ensured that the development of discursive institutions was at the heart of the monitoring effort. One illustration of this commitment to the essentially process nature of monitoring concerns the bankruptcy of enterprises before or following reorganisation. As fiscal constraints started to bite in Russia, and bankruptcy became a live issue for the rural sector, then bankruptcy became a priority for project strategy and monitoring. Some participants had clear ideas about what should be monitored and how this should be undertaken. Nemtsov, for example, wanted comparisons between financially 'strong' and 'weak' farms, some of which had 'split' and some of which had not changed their structure. Others stakeholders could only broadly agree that issues of potential and actual bankruptcy were important and that they should be monitored and reviewed from time to time. A procedure for dealing with bankrupt enterprises could never be agreed but the issue continued to be discussed and was to an extent resolved in the agreement not to undertake a special study. According to Russian officials, by the end of 1995 nearly 70 per cent of the project farms were 'technically bankrupt'. In practice, the local authorities found ways of keeping them afloat and actually stepped in to prevent the project piloting a bankruptcy process on one farm because of the potential political damage it could do to the project. The project staff questioned how useful yield and financial information could be when managers and others were facing 'bankruptcy'.

The intimate relationship between attempts to change the pattern of economic transactions and sustained project discussions aimed at developing a common understanding of the world has made it possible to identify, and gain acceptance for, a main core of 'impact' propositions to be tested with information from the monitoring system. These propositions concern *both* the farms' responses to market forces and the quality of the reform process. The key ideas to be assessed are that:

- farm management and farm owners change their behaviour in response to market pressures;
- changes in ownership bring about changes in labour incentives and work behaviour;
- privatisation provides the basis for an efficient market in land and other assets and factors;
- the privatisation process is legal, fair and understood by those involved and does not create undue feelings of uncertainty and dissatisfaction;
- legality, fairness, and transparency can be maintained through lower cost approaches than those used to initiate the process.

The initial impact monitoring focused on five pilot farms. It was carried out by the Russian professionals on the project and by various students from a variety of disciplines. There were two main teams divided by discipline (an economic monitoring team and a social monitoring team). In addition to the monitoring teams a great deal of information about impact was coming through the day to day project activities on the farms, for example from the post-privatisation business consultants and from the lawyers helping to sort out legal problems.

The economic monitoring team divided the types of information they collected into three categories. First, economic analysis which collects data about physical and financial indicators on the farm. Second, legal and accounting activities which describe ownership structures and other legal relationships and note changes in them. Third, confidential/narrative information which the team hears or gathers on farm and which is generally treated in confidence and not released widely. In addition, to allow aggregation and comparisons with non-project farms, the team uses official statistics submitted to the Department of Agriculture.

There have been several problems with the data including the short time period for analysis (the first privatisation was in early 1994), lack of a suitable base case or control group and the inability to separate out macro-economic effects from specific privatisation effects and the problems of obtaining accurate and non-confidential information – especially about 'grey economy' activities.

There are similar problems facing the social team. When the project started those involved felt that it was not sensible to have a base case or baseline survey. Comparison with past non-privatised performance was considered not to be useful because so many other parameters were changing very fast in Russia and were difficult to control for. Similarly control farms were not identified – partly because of the overriding focus on finding out what changes the people on the privatised farms had experienced. People's comments and observations of change were considered to be more useful – especially for management information purposes and feedback into project design – than any comparison with a formal base case. Resources were stretched and a base-case survey or group of control farms seemed a luxury. Much of the social data is collected using Participatory Appraisal Techniques and from a sociological survey administered to a sample of reorganised farms and those not reorganised.

But, as the project has moved on, this lack of a control group of farms has been a constraint in reaching conclusions, especially in terms of isolating the impact of privatisation compared with other changes in the agricultural environment (liberalisation of marketing, etc.). Comparisons have mainly been handled through using formal statistics still gathered by local authorities but this in itself has posed problems because of the problems of size (i.e. it is difficult to compare the results from several smaller farming units with old style huge collective farms – there are aggregation issues).

The base-case issue has also been very difficult for the social monitoring team. Many people are blaming privatisation for a reduction in the quality of social services when in fact this seems to be a common trend across rural Russia. In some circumstances control farms have been used to examine particular issues but there are problems in obtaining reliable information. The control farms do not trust project staff at all – they have no relationship with them and often no interest in the issues which they are not part of – they often pose political resistance to the teams (remember the control farms will normally have chosen not to go ahead with privatisation). And so the quality of information obtained from the controls is very poor even using PRA techniques. Some success has been had in working with control regions using focus groups run by an independent PR company. The neutrality of the company has helped the focus groups to unpack attitudes and blockages to privatisation.

Both the economic and the social teams use students trained in PRA. An important feature of the project is that the PRA trained people (who had to be trained anyway to conduct the information campaigns as part of the project) were used to gather hard financial and economic information. Using a basic set of issues and key questions prepared by the economists, the information that was gathered was considered to be much more realistic than the financial returns the farms were making to the local authorities.

PRA was successful in gathering information about grey zone and black market activities and in breaking down the existing distrust of officials and use of information. But problems of confidentiality remained. Project monitors have to be very sensitive to confidentiality. Exposing financial information to the local press etc. can make farms very vulnerable to increased taxation and sometimes to the Mafia.

The project has been growing. New regions have been adopting farm privatisation and doing so with a much lower level of project support. One of the aims of the present phase is to show that less involvement by project staff and more management by Russian organisations can still lead to a satisfactory farm privatisation impact, particularly in terms of legality and fairness. Mobile teams, a 'hot-line' and other systems have been put in place to support the process.

As the project grows in size and the number of privatised farms involved in the process increases, the demand for information increases. The intensive monitoring of process and impacts on the original pilot farms cannot be replicated across all farms and regions involved in the project without unsupportable costs. The project must, therefore, evolve a modified monitoring approach. The aim is to continue process and impact monitoring methods used successfully in the previous stages while also trying to evolve a system which is local and includes an integral quality assurance mechanism.

The following existing features will continue into the new system:

- a flexible process with the ability to respond to different emerging priorities and to expose new problems or successes as they arise;
- asking farm people to identify change rather than conducting expensive base lines and controls;
- a mixture of survey and intensive 'case study' monitoring on the initial pilot farms;
- topping up information derived from operational work with intensive monitoring, at annual or twice yearly intervals, of specific processes using a mixture of survey and PRA techniques.

This combination of methods accepts that process monitoring and impact monitoring are hard to separate. If the project monitors the pilot farms, for example, it monitors the on-farm changes in terms of farm efficiency, management behaviour and good governance, level of social service provision, wage payments, decision-making, unemployment etc. The use of qualitative and quantitative methods ensures that a large array of information is collected which can be used to address all these concerns and makes it difficult to draw a realistic borderline between process and impact.

The need to add some kind of in-built, local control of quality into the monitoring system emerged as project staff realised that relying on continued intensive (and expatriate) monitoring would just be too expensive as the process was transferred to Russian management. The project monitors understand the need to provide a combination of informed, hands-on management and independent monitoring. The challenges to the project, now being addressed, are how to gain Russian stakeholders' acceptance of the need for balance between participation in the process and autonomy in impact assessment; and how to develop a mechanism or select an institution whose authority in achieving a reasonable balance will be widely accepted.

Although there is as yet no final resolution of this matter, it is clear that the problem is both a general one facing monitoring systems and also a specific problem for monitoring development projects in contemporary Russia. First, there is the general question of timing and the differences between early and later stages of project implementation. In the early stages of a project intensive process monitoring is necessary to fine-tune the approach. In the later stages, however, donors and/or project opponents may propose more 'neutral' monitoring to allay suspicions that vested interests are influencing the results and assessments. Second, as implementation has been transferred to Russian management, and lower-cost replication methods used, the responsibility for guiding the reform process lies increasingly with the local administration. A Russian foundation has become the 'Russian home' of the project and is responsible for using project funds to assist the Russian regions in implementing further land privatisation activities. The project must therefore establish the extent to which 'the Russian home' should again blur the distinction between process and impact to achieve more effective local management or should establish independent, impact monitoring.

Finally, the direction and definition of 'development' in transitional Russia is a battleground of controversy, contending views and national pride. Policy changes or reform activities involving cooperation with Western organisations are scrutinised especially carefully and are treated with suspicion. Russians see themselves as nationals of a country that symbolised a system. They are nationals of a once very influential country, formerly at the forefront of social progress but now, following the offence to pride which started in 1991 with fourteen states cut from Soviet territory, subject to the advice and wishes of Western experts. So a debate about technical project planning issues such as monitoring can readily become part of the debate about recreating a distinctively Russian way of development. Some stakeholders will argue strongly for 'independent' results. Other stakeholders will ask for more, short-term local development 'action' and less foreign interference in the meaning of fairness, legality and cost-effectiveness in the distinctive Russian context.

Policy reform in the Indian forestry sector

Background

Rights to land and its products and the balance between the rights held by the state and by local communities and residents are also the keys to policy reform in the Indian forestry sector. Over 75 per cent of the Indian population of nearly one billion live in rural areas and are predominantly dependent on natural resources for their livelihoods. A key natural resource is land devoted to woody vegetation and its associated pastures and wildlife – in short 'forests' whether these are open savannah lands or dense tropical rainforest. The products of this land type can be used for fuelwood, for pasture, timber and poles, for a great variety of non-timber forest products such as mosses, berries and fruits of use in food and medicines. Forests are central to many rural people's lives – their wood, leaves and fruit providing essential building and cultivation equipment, paper, plates, grazing, food and medicine and the protection of their watershed and farming systems.

Forestry is a major policy area because of the need to regulate over-use of dwindling natural resources and to create resource sustainability for a growing population which includes many of the world's poorest households and individuals. Forest policy also has important ethnic, cultural and poverty reduction implications in India. One-third of the world's tribal people live in India. Most of them live in the broad strip of forests transecting the sub-continent from Gujarat to Bengal. In 1991, the majority of Indian's sixty million tribal citizens lived inside or near the *sal* (shorea robusta), teak and acacia forests of the central region. They include among their numbers some of the world's poorest social groups. Isolated, and with minimal political influence, they have little effective access to education, healthcare and income. Agricultural and industrial pressures on the forests are also degrading the resource base on which they depend for their livelihoods and subsistence. Forest policy is thus simultaneously a major contributor to ethnic and tribal policy in India. It also has important environmental implications since the region is the source of some of India's largest river systems and the forests of the region contribute to the climatic and hydrological regimes of the nation and to many of its industrial products.

Forests and forest lands in India are state controlled and administered by state forest departments. Policy on forest environmental protection and use are also key national concerns. Access to the land and the forest products is intensely sought by timber companies, rural people and by water and mineral resource development projects. These many competing uses and claims on the forests are arbitrated and policed, at least in theory, by the forest departments. Governments in search of revenues have encouraged intensive timber exploitation in the past. Population growth and economic change flood the forests with migrants searching for farmland and grazing and saleable products (for example, large leaves to stitch together to make 'plates' for food). Politicians and government officials have sought in the past to have forest lands transferred to their control, to reclassify the land for resettlement projects, and to develop the lands for plantations, private fishing or industrial use. Environmentalists advocate the 'claims' of the forest wildlife, its ecology and biodiversity. Social scientists and local representatives point to the needs of the indigenous tribal and forest communities living in or near the forest.

The forest departments and policy reform

Increasingly, it is recognised that the forest departments lack the capacity, staff or resources to successfully arbitrate these many claims and to conserve the forest's resources while doing so. Moreover, these forest agencies have given little attention to understanding the social problems related to their field operations; their arbitration tends to favour those with power, influence and economic voice. Palit (1996: 218) estimates that within these large forest bureaucracies there are somewhat more than 120,000 forest staff involved in management activities and responsible for safeguarding and controlling use of about 23 per cent of India's land area and its forest resources. Most of these individual management staff have heavy, routine administrative duties and have little or no time for field work. Lacking either the staff, or the planning tools and incentives to respond to diverse local conditions and needs, forest management staff rely on standardised technical procedures and a culture of 'target' achievements and legal-bureaucratic supervision of junior staff. These forest bureaucracies now also have long corporate traditions of operational working, recruitment, training, service histories and technical expertise. They have abundant knowledge of silviculture for timber production, of soil and water conservation planning, and forest law enforcement. The many competing claims for access from the rich and influential, the need of the forest departments to protect their institutional traditions and territory, and the increasing number and scope of routine administrative procedures are superimposed on the biodiversity, social and cultural variation, and locally specific needs of plant, animal and human communities. This combination of political pressure and proud forest service standardisation with complex local needs creates a challenging field for both understanding and the design of measures for policy reform. Change in the norms and standards which govern institutional operations and in the policy context is very difficult, in part because the implications are often so complex and they involve so many stakeholders and interest groups.

The degree of complexity perceived in the process in part depends on theoretical perspectives and in part on the nature and history of the sector. For those who believe that public policy is ultimately settled by the decisions and penstroke of a Minister, the problems of forest policy in India will suggest a lack of political will and purpose. A more 'incrementalist' perspective on policy reform will see that the detailed shape of a sector's public institutions and agencies is itself always a major factor in determining the norms and standards governing official behaviour and understanding. Policy reform in the Indian forest sector necessarily requires, in this perspective, detailed attention to the policy and institutional process and to the prospects for the incremental resolution of successive problems. As a sector, moreover, forestry does have its own distinctive features and problems which reinforce the need to consider the institutional process. The major Indian bureaucracies providing routine and mass civil administration and essential individual needs such as health, water, security and education have had to devise ways to ensure upward communication channels for information on community matters. The historical core of forest departments' work, however, has been to manage publicly-owned landed estates through a combination of enforcement, industrial liaison and technical expertise. 'For decades, forest departments have acted as the sole custodians of vast territories: in essence as "state zamindars"' (Poffenberger and Singh 1996: 79). Attention to the social problems related to field operations has been of secondary importance. Moreover, as the most recently created of the three major all-India services – the civil (IAS), police (IPS), and forest (IFS) administrations – the IFS lacks something of the general power and the prestige of the IAS. These features can make forest departments inward-looking, jealous of their existing powers and self-consciously 'scientific' to the detriment of information on community relations and local and social objectives. There are, of course, a growing number of instances of reform and sustained individual innovation with considerable potential for meeting social and environmental objectives. Nonetheless, as in the project to change Russian agricultural institutions reported in this chapter, the incorporation and institutionalisation of instances of innovation and reform into the nation's generally over-extended bureaucracies dealing with land management are a great challenge for policy and institutional analysts.

International donors can provide diagnostic help from a variety of sources and other support to government to enable forest agencies to develop new programmes and the space for learning lessons from innovations and incremental reform. This section of the chapter will concentrate on the policy and institutional experiments in three states drawn from three forest zones in India – Orissa in the central tribal belt, Himachal Pradesh in the Himalayan zone and the Western Ghats forest in Karnataka in southern India. Features of the three zones and aspects of the forest management in each of the three states are summarised by Poffenberger, McGean and Khare (1996: 28–46). The reader is referred to this source for a discussion of regional patterns of local environmental activism and the character and state of the forests. Although these three states and zones differ in many respects – for example, in terms of the incidence of poverty, the concentration of tribal populations, NGO numbers and activism, the speed and ease of natural forest regeneration, the type of forest and the extent of existing forest cover – they do share common features. In each case, there is a close relationship between the local people and the forest. In each case a major forest department is beginning to find a way to work with local communities and undertaking measures of decentralisation and policy reform on behalf of 'forestry for people'.

In each case a major forest department has helped the Government of India secure bilateral technical assistance grants on its behalf by agreeing that it will implement participatory forestry using a combination of all-India policy and regulations, state law, international best practice and some measure of decentralised decision-making. In each case the international donor has embarked on a programme of support for change in the forest bureaucracies – armed with confidence that institutional change was necessary, and a hope that variants of 'process' planning would identify the scope for institutional change and allow lessons to be learnt and implemented.

Process planning and documentation prospects

There were very few written guides or analyses available to guide the design of the institutional changes needed. Poffenberger's (1990a) chapter 'Facilitating Change in Forest Bureaucracies' was perhaps the only sure-footed guide available at the time the projects were being prepared. The disappointments experienced in at least one of the three case studies are mainly the result of not following, or being unable to follow, the lessons Poffenberger had learnt in Southeast Asia. The 'process planning' framework Poffenberger is recommending is relatively simple: a first phase of diagnostic 'action research'; a second phase spent integrating the new procedures generated by the diagnostic research into agreed planning procedures; a final, third phase of expansion for the new approaches; recognition that the required changes would probably take a decade or more; special care taken when judging the optimum pace of the process; and the extensive use of process research teams/working groups.

Poffenberger's (1990a) analysis of Southeast Asian forest bureaucracy concentrates on the constraints and opportunities faced by working groups/research teams when trying to achieve change. His account is not a detailed documentation of institutional process in one or more departments but it is based on field work and does convey the experience of learning process projects within forest administration. Since the time it was written, Palit (1996) has added a notable commentary on what forest bureaucracies should do; Hobley (1996b: 236–8) provides a significant analytical list of overt and covert institutional patterns. Otherwise, documentation of the institutional change process in forest departments remains relatively meagre. The published record is no greater than the sum of the references just cited. Unpublished consultancy reports could provide further valuable information but they have their own conventions which make it difficult to report the process in detail and the full set of lessons learned. Sarin (cited in Hobley 1996b: 239) identifies what can be, or is being, undertaken in process monitoring but few results or conclusions are so far identified.

These thoughts on process change and its monitoring in the three chosen states are offered in the absence of detailed process documentation. Process monitoring is an essential part of any learning process project because it is crucial to expose and resolve the often contradictory visions of social integration and benefit that are held by different project stakeholders. The aim of planners who are preparing process projects or programmes is to achieve a combination of commitment to, and coherence in, the aims of the project. Commitment or 'ownership' is needed to ensure that project stakeholders continue to support the project's aims once the aid agreement has been signed. Coherence in the project's aims is also required to ensure that it can be monitored for effectiveness and impact. Unfortunately, these two process project aims of 'ownership' and 'coherence' may be incompatible. 'Ownership' is usually achieved through joint planning and the acceptance of a diversity of views, perspectives, contradictions and shifting compromises. Commitment or 'vision' is also usually achieved through the articulation of aims by diagnostic teams ; but 'vision' also usually implies a homogenisation of views through joint work, or the domination of one set of views which is then given a special authority by adoption as the project's 'vision'. A formal coherence may be achieved but the 'vision' usually privileges the views of one set of stakeholders and marginalises others.

If senior management can create a moral authority for a new vision of the institution then the way ahead in the process project can be straightforward. Often, however, and in each of the three forest case studies being dealt with here, senior management has found it much harder to agree the direction of change or to create agreement and commitment outside the senior management group. In these cases, insistence on 'vision' is unsustainable and threatening to the project's eventual success. It is usually more feasible to discuss the aims and hopes of the project in more workaday, less encompassing, terms and to set up working groups or diagnostic research teams which can propose new operational norms and standards as benchmarks of best practice. The piecemeal engineering of these benchmarks may, ultimately, more easily anchor institutional innovation and change than general discussions of aims and purpose. The role of the process monitor is to clarify the shape of these different views held about the project and institution, to access alternative examples of good practice, and to indicate the probable consequences of taking one path or another.

A forest of visions and institutional discourse

In each of the three forestry projects dealt with here, there have been essentially contradictory perspectives and visions about the nature and role of participatory forestry, even though there has been little disagreement about the need for forest departments to create better public relations or for the greater involvement of local communities in forest conservation. All forest department stakeholders are aware that forest departments have public images that could be greatly enhanced. Forest officials have been accused of being prisoners of timber and pulpwood industry interests, prisoners of local rural elites, tolerant of the petty corruption of forest guards etc. and stakeholders are aware that these images threaten or even undermine many new local initiatives and have burdensome political repercussions. It is not hard, then, to achieve agreement in favour of 'participatory forestry'. The difficulties start when actual forestry activities and operations involving participation are discussed.

Despite superficial agreements about image and public relations, the contrasted visions of society and social integration and exclusion which are held by project stakeholders have many different consequences at field level. To distinguish these visions or discourses of participatory forestry we can term them F2, F3, etc. The visions are used by project participants to explain the shifting and contested process of participatory forestry to themselves and others. They serve, as explanations, to reduce the complexity of the process and the impact of cognitive dissonance brought about by uncertainty and change.

F2 is *Friendship Forestry* (or, in the Indian context, *'Folded-hands' Forestry*). In this simple model, participatory forestry mainly requires forest officials to smile, greet villagers respectfully and encourage dialogue and consultation. It is expected that villagers will be pleased to meet friendly foresters and will agree to follow the forest department's instructions. The vision is – change your behaviour when dealing with villagers and your work as an official will be made more comfortable.

F3 is *Friendly Freemarket Forestry* or 'the human face of sectoral adjustment for forestry'. True friendship and sustainable joint forest partnerships require, in this model, that the forest sector continues to be friendly but does so by first getting its prices right. Forest departments should start by removing their large subsidies to the timber and woodpulp industries and to

177

urban fuelwood consumers; these consumers are assisted with direct subsidies and to start plantations whereas the poorest consumers of the forest resource – the headloaders of fuelwood and forest floor graziers, etc. – are merely punished and given no investment help. The vision is – get the policy environment right and the rest is relatively routine and unproblematic administration.

F4 is *Forestry's Flourishing Fabian Fable* or a vision of social justice achieved through staff training and the adoption of populist methodologies at operational levels. It is a fable because it assumes, with much evidence to the contrary, that institutional reform will follow better training and the deployment of technical improvements in data gathering and in decentralised management. It is flourishing because of the popularity of Joint Forest Management which is treated as a largely unproblematic vision of decentralised empowerment for disadvantaged forest dwellers and their near neighbours. The vision needs only sincere forest officials trained in PRA and related data gathering techniques. It is Fabian because the Fabian Society – a part of the British Labour Movement – had a largely unproblematised view of the social administration that was necessary to introduce the welfare state and nationalisation in Britain. It was assumed that social justice could evolve through political will and the routine administration of gradual policy changes by existing civil service and local government machinery. That the machinery of government and the institutional allocation of public goods had the potential to be every bit as regressive as the market and as distorting in its welfare outcomes, was not appreciated by the Fabians. The vision is – commit the institution's staff to social justice through training and the use of principles of fair, universal coverage.

Since the time of the innocent Fabians, however, experience and analyses of the 'bureaucratic phenomenon' have shown us that, even if training and attitude are excellent, 'individual decisions do not, in a world of institutions, necessarily add up to rational institutions and rational systems' (Peattie 1994: 124). Part of the problematic nature of administration lies in the relation between information and power, or between principal and agent. Michel Crozier (1964) demonstrated for us, in the case of French administrative systems and more generally for example, that if every administrator acts in the way that is most rational for him or her, decision-making is increasingly pushed upwards while information is retained downwards. The result is a system based on a major separation between the power and information necessary for sensible decision-making. Forest departments in India feature many of the features of the 'bureaucratic phenomenon' that Crozier describes for France.

Furthermore, the pathologies of administrative systems are not usually subjugated with earnest talk of social justice. Development policy discourse and its institutional agencies add a further layer of complexity, bias and intractability. Apthorpe (1986: 384) dissects its special languages to show that the system 'construct[s] series of terminologies with the aim that exposure should be avoided at all costs – but in the name, however, of openness, science, impartiality'. We see, in the utterances of development policy, this encoding of conflict, culpability and institutional pathos under the guise of rational, 'neutral', 'fair', 'consultative' reporting. In addition we also see the distinctive style of the international development discourse – that it aims to be donative. It promises to give, to deliver.

> It is solution-side utterance, a form of teleological willing. Its statement of the problems is very much determined by the expected nature of their solutions: thus anticipation of solutions comes before statement of the problems . . . Development policy discourse, so that it may move as quickly as possible to solutions, prefers to waste very little time on detailed diagnosis of problems, as far as possible to distance policy solution from problem genesis. It is, as it were, simpler and quicker just to aver that 'a crisis' is at hand . . . Hard demonstrations of diagnosis, of actual pathology and actual links between pathology, remedy and prognosis, are absent.
>
> (Apthorpe 1986: 386)

Participatory forestry has remained reasonably distant from these conundrums of information, power, terminology and discourse within development agencies – as the literature which generates the puzzles would predict. The view of state action underlying 'JFM' is, by and large, Fabian and uncomplicated – arguing for more 'state support' and donation – or, less frequently, neo-liberal and equally uncomplicated – arguing that the state should be downsized and 'kept away' from local community resource management. The policy and power constraints and what Hobley (1996a: 221) terms the 'hidden' aspects of forest sector institutions – for example, rent-seeking behaviour and the use of office to create ways of supplementing meagre incomes – are largely ignored.

Each of the three 'visions' of participatory forestry held by forest departments in India is held in different degrees and by different groups. Each discourse of participation simplifies the project process because all of them are applicable to a degree. There *is* a need for foresters to improve their styles of interacting with villagers; there *is* a need to remove distorting subsidies; and there *is* a need to introduce improvements to staff attitudes through training and rational, universal procedures adopted for the existing administrative machinery. Taken singly or even together, however, they do not adequately explain the difficulties of policy making and implementation or its monitoring. Poffenberger sensed this when he wrote that 'assisting government bureaucracies to make systematic transitions from being resource managers to helping communities develop resource management capabilities is a complex process' (1990a: 117).

Part of the complexity of the process to which Poffenberger refers is that 'the hidden institution' is not in fact very well hidden. It is tacit in official development policy discourse but well known to all primary stakeholders and the officials themselves. Vanaik writes that detailed studies of state action in India will confirm 'the immensely superior access that, for example, business and rich farmers have to the Indian state compared to labour and the poor peasantry; or, the ways in which they create obligations through campaign contributions, corruption and favours to superior bureaucrats and politicians' (1990: 15). It is because of the inherent difficulties in enforcing universal rules of protection and controlled access to common property forests that Blair (1996), in a recent review of common property forest management, casts doubt on management arrangements in which the state takes the lead. In Blair's analysis it is only local groups in south Asia's villages adjacent to the forest who have the best chance of managing the collective property forests. Local and central state agencies have no real incentive or ability to protect the forest even when they, formally, own it; and they are heavily permeated by the overt and tacit claims of business and local elites and political influentials. The learning process approach states that this complexity of claims, weak incentives and the multiple visions of participation and forestry cannot be avoided – they can be reformed after careful diagnostic research and monitoring in order to identify lessons and to suggest ways to best incorporate possible reforms in institutional practice.

Gains and reversals in the learning process

The three states studied have had varying degrees of success with the introduction of new arrangements for community managed forest resources and with the monitoring of lessons learned from the changes. In each state there have been substantial gains – and many reversals. In the eastern, central tribal belt case study, the gains in social authority in forest management have been very significant. One context is the extent of forest dependence and the rate of forest degradation. Orissa is one of India's most important forest and tribal states, and one of its poorest. Thirty-eight per cent of the state's land is designated as state forest and there are many tribal groups living on forest land or nearby. Commercial logging, fuelwood extraction by individuals with headloads of wood, and overgrazing of the forest floor have led to widespread deforestation and to the diminished flow of forest products which has been critical for the livelihoods of many tribal communities. One estimate gives a 10 per cent decline in forest cover in Orissa between 1983 and 1987 (Poffenberger *et al.* 1996: 35). Even allowing for gross miscalculations and inaccurate information it is likely that the loss of forest cover has indeed been dramatic in recent years. The other context for community forest control is the continued or renewed vitality of village management systems for natural resources. Hamlets are frequently relatively homogeneous in tribal and caste terms and have traditionally been involved in a range of common property governance activities including land and forest protection and the maintenance of temples and ponds for water supply and fishing. In response to resource depletion and using tradition-based patterns of resource management, many communities in Orissa began to protect local forests in the 1970s. 'By the late 1980s, an estimated three to four thousand communities had established control over about 10 per cent of the state's reserve, demarcated and undemarcated forests, covering some 572,000 hectares' (Poffenberger *et al.* 1996: 34). By the end of 1993, about 27 per cent of Orissa's state forests were under community control (Poffenberger *et al.* 1996: 35). Kant *et al.* (1991) describe some of the successful cases; these cases preceded the enthusiasm for 'JFM' and have themselves encouraged 'JFM' solutions in other states.

The role of the forest department in these developments is often unclear. Often *F4* assumptions dominate with sometimes crippling consequences. In late 1988, for example, the government of Orissa passed the Indian nation's first forest policy resolution endorsing community management. The announcement assumed that the state government would take the lead in consolidating local achievements and amplify them by making community management gains universal in coverage and benefit through official action and endorsement. A new government scheme defined ideal groups and their institutional structures and ordered the forest department to form 5,000 new forest protection committees that year. The majority of these new groups failed and the traditional groups went unrecognised. An opportunity to simply recognise and then define a framework of support for existing local forest-protection systems was lost.

Nonetheless, despite false starts, there are active examples of very supportive relationships in which the forest department has effectively delegated the protection of forest resources to hamlets or villages. It continues to provide valuable support to the communities in their relationship to other state government agencies and frees the forest officials to concentrate their efforts on the protection of the Reserve Forests. SIDA is assisting the Orissa forest department to plan its future role and activities in the light of changing demands and contexts and is using a combination of forest department diagnostic teams and a light leavening of external, Indian and European, advisers to introduce new ideas and methods. There is a strong commitment to the learning process approach and funding is linked firmly to progress in achieving process objectives (Rew *et al.* 1995).

The combination of visibly increasing forest degradation, high tribal and other community dependence on the forest, and successful forest protection initiatives have simplified the choices concerning 'vision' for the forest department. There is a weakened regard for *F2* – it has been tried, tested and superseded. The strength of local forest protection has discouraged belief in the conventional (*F4*) assumption that the forest department will simply design and administer beneficial regulations on all forest users' behalf: the links to commercial interests, the difficult access relationships, biases and processing difficulties inherent in government machinery are now more fully appreciated. *F3* continues as a favourite diagnosis of visiting policy analysts and of some, but very few, NGOs. Local officials continue to recognise the continuing importance of commercial influences on the forest sector and appreciate that the *F3* vision of changes in the system of subsidies for fuelwood and timber extraction and in the monopoly pricing of non timber forest products (NTFPs) – on which many tribals depend for off-season income – is very difficult to implement.

Process monitoring through social science facilitation and the interpretation of institutional action and policy initiatives has been in place and guides the lessons being learned about the composition and nature of working groups and the form of their analyses and about the feasibility and pace of change. As in the Russian land privatisation case, there is always the possibility that 'monitoring' is taken to mean a quantitative assessment of outputs and impacts, while it is often difficult for the participants to distinguish process monitoring from management with a light touch. In neither the Russian nor the Orissa cases are the existing arrangements for process monitoring likely to lead to a large body of written reflections on the project process or on the general lessons to be learned for process projects elsewhere.

Sood (1996: 12–14) reflects on the process of forest management change in Himachal Pradesh. He recognises the importance of timing and the pace of process change, extensive consultation with stakeholders and of non-prescriptive support for process change. He also records the critical sensitivities of the problems and procedures being analysed and discussed. He records the verdict of an Himachal Pradesh NGO leader (1996: 16) who cautioned foresters that by decentralising decision-making to users they were playing with fire ('*Aap aag ke nazdeek ja rahe hain*'). The Himachal Pradesh case also identifies (Sood 1996: 17) that forest department service conditions and incentives for junior staff may threaten significant change in attitudes and procedures: 'Bleak promotional prospects and limited monetary incentives reduce the motivation of FD staff to adopt new approaches'.

There have also been significant gains and changes in the Karnataka project – the key sources are Locke (1995) and Luthra (1996). New procedures for site-specific planning have been adopted and much experience has been gained in forest policy and planning at Circle level. Difficulties have also been experienced in the 'ownership' of the process and its scope and aims and in the extent to which different visions of the process continue to exist and vie for attention.

Despite these gains, all three states have experienced many reversals. It would be incorrect to single out any one of the states and to describe their particular difficulties. They have all had periods of extreme tension in trying to decide the pace and direction of change. If the pace is too hurried, the process of change stops and existing methods are used uncritically – either from preference and lack of challenge, or because there is a lack of new alternatives. If the pace of change is too slow, and there is the appearance of endless discussion without field work or new initiatives, frustration sets in and agreements begin to break down.

In the paragraphs which follow, the particular experiences of one of the three states have been used as a base from which to characterise the difficulties and challenges of a process approach and the hazardous setting it creates for project monitoring. Certain of the specific features of the case have been disguised to ensure that it provides lessons for *all* states introducing Joint Forest Management and leads to more socially responsive forest services, not to defensive reactions to perceived criticism. Thus, although the circumstances of the lessons which have been learned are specific, the intention is to make them general in their application.

One unfortunate aspect of the case, which provides a strong cautionary note for continuity between process project design and the early stages of implementation, is that an essential part of the process was forgotten, compromised or inadequately delegated once the project started – certainly it did not work effectively. While the project was designed through consultation and through the use of facilitators and consultants who were independent of the donor and executive agency teamwork, at the implementation stage it was decided to recruit a specialist in process projects but to recruit him as a member of the donor's own administrative office in the field and to give him budgetary and office administration duties as well. Inevitably, the debate between the forest department and the process specialist was seen (by the forest department, the donor and by technical consultants to the project) as the continuation of negotiations between a representative of the donor and forest department officials. The donor's strongest vision of institutional reform (largely F3) was never made fully explicit until later stages because of the need to develop 'ownership' on the part of the forest department through the articulation of specific needs in joint planning sessions. Many officials in the forest department really wanted – in terms of the 'hidden', or better 'tacit', rent-seeking agenda of the institution – large-scale, technical solutions to implement. It was, in manpower and revenue terms, a strong forest department and wanted more schemes to implement and money to disburse. It was not about to have its remit seriously questioned – few were really interested in changing the tacit institutional behaviours. Their 'ownership' was restricted to the aim of environmental conservation and to a limited view of social provision in the sense of the basic forest product needs which a forest landlord could extract and give or sell cheaply.

The balance between ownership and vision is at the core of the problem of deductive process and process diversity referred to in the introduction. The process project planner must find ways to dissolve the conflicts between competing visions and then to resolve them in terms of new action research and policy diagnosis agendas. This dissolving and resolving is often best achieved through new action research results. There was no shortage of interest in research and research degrees on the part of the senior forest department managers – and these may have been a necessary incentive. It was in fact some of the donor representatives and consultants active in the preparation period who were less sympathetic to research agendas, preferring solution-side diagnosis rather than messy, meaningful and contestable reporting. They supported action research but wanted this very firmly tied to the new deductive, planning machinery to be established by the project. Unfortunately, new planning posts which were established proved difficult to fill with appropriate staff so that a programme of free-ranging action research never developed at either decentralised or centralised levels. Some operational experimentation was tried and was successful but the Circle action research response was disappointing; and at Headquarters the critical diagnostic teams and research programmes never emerged to signal process innovation.

The lack of action research and committed planning staff was part of the problem. The relative lack of recruits to decentralised action planning can be explained by the lack of incentives in research and planning work compared to the 'incentives' or rents in the 'tacit' institution. The main constraint on process change, however, was the corporate strength of this particular forest department and its largely *F2* corporate vision. It wanted to incorporate lessons on participatory forestry and to repair its public relations image with the public and NGOs. The wish to transform the edges of its working life but not the core is perhaps not unusual. But it led the forest department concerned to demand the incorporation of foresters as institutional organisers, however unlikely this appeared. It certainly had very little sympathy with the social development consultant's recommendation that special staff with an NGO background should be recruited as Community or Institutional Organisers, either as individuals or as part of a link organisation, to serve on multi-member Joint Forest Planning teams. The aim of the recommendation had been to establish diagnostic teams that would include a range of foresters and community organisers, legitimate a variety of visions at site, range and Circle levels and create the basis for social authority in forest management decisions. It was, however, especially difficult to get acceptance of the idea that 'officers' were by definition and label not likely to be good at community facilitation.

Social science finds it hard to predict outcomes – so much of social life is contingent and contextual. In this case, however, it was not hard to predict, at the time, that officials, and especially forest 'officers', whether with or without uniform, would make poor organisers of local institutions. Their background of enforcement and official action is uncongenial to the task. As officials, they also cannot help being seen, and acting, as agents of state authority. Where, as in Indian forest administration, there has been a history of difficult state to community relations it is impossible for 'officers' to create social authority for villagers even if they are accomplished performers in the style of F2 – and not all were. Uphoff's (1992) account of participatory learning in the Gal Oya irrigation scheme in Sri Lanka shows, convincingly, that specially recruited institutional organisers can break the downward spiral in the vicious circle of low social authority in the community and distrust of government.

The drive for coherence also led to a sort of instant validation that was premature for the development of the process. The state legislature introduced a Joint Forest Management regulation very early in the project life and before there were any results from the project planning process on which to build. The actual regulations took a predictable form and, although not perhaps optimum, as a set were broadly suitable and hopefully open to subsequent amendment. The specific problem with pen-stroke validation is that the process element in the project was made immediately problematic. 'Officers' were the appropriate instruments to implement a strategic government regulation. The need for foresters to serve as official proxies for non-official community organisers and regional poverty analysts was made superfluous at the very time they were made responsible for the registration of new official villages committees and the micro-planning of adjacent degraded forest land.

In this generalised case, process monitoring was seen, by the donor, as essential to the success of the project. It proved impossible to institutionalise the role for a number of reasons. There were changes in the personnel responsible for this function at the many stages of preparation, appraisal, mobilisation and full implementation. Independent process monitoring was confused or conflated with the donor's social development monitoring. Tensions between the compromises needed to achieve stakeholder 'ownership' and the drive for coherence in design needed for accountability reasons were never fully recognised and resolved. There were widely contrasted visions and fears about 'research' and its relevance. One aspect of process monitoring was delegated to an independent contractor but this delegation was then withdrawn; and the justification for process monitoring was not fully explained to the forest department – the discussions centred on training, new skills, new roles and management improvements rather than on a system of process change.

The difficulties in explaining the concept of the process project approach, and in finding locally available process monitors and researchers are similar to the difficulties in achieving high quality social science field work reported for India (see Breman 1997: 111). The sanskritic tradition in scholarship places an emphasis on the study and use of textual authorities and regula-

185

tions and leads to a lesser regard for empirical field work; when it is under-taken its quality is often not the best. Beteille (1996: 234) traces the relatively lower quality of Indian field work to the repeated but much shorter field visits and the scholar's tendency to dispense with learning new languages and to make other shortcuts. There are many implications of this difference in approach and in results. First, there are much fewer local role models for aspiring process monitors. Second, it is harder to explain the need and role to those not previously exposed, even indirectly, to social science field work traditions. Technical assistance agencies such as SIDA and ODA (now DfID) have been exposed to social development thinking and anthropological analysis for a decade or more and have learned its benefits and how to incorporate the perspectives in their projects. The use of social science in Indian policy and planning is very rare and even those anthropologists employed by national planning agencies find it difficult to get their discipline acknowledged and utilised. Third, the use of social and cultural analysis to address management issues is a pre-requisite and universal in process projects for international development; it is a clear but minority component in British management science teaching and research; and is very rare indeed or non-existent in Indian approaches to management and development. Taken together this combination of factors implies that the process monitor will most likely be an expatriate. If, as in the case of the Himachal Pradesh case, a monitor is a serving Indian officer he will need special talents, a protected role and a keen sense of what can be reported and to whom. The few Indian process monitors who will emerge are perhaps best placed to give initial shape to the methods and focus of any *F5* level of analysis – that is, a documentation and analysis of the overall complexity of a project process and its combination of resolution for competing claims and the achievement of coherence for accountability reasons.

Lessons for process monitoring and research

It's good to talk

The key challenge for the Russian land privatisation and Indian forestry projects has been to establish 'discursive institutions' which could review the outcome of attempts to change the pattern of economic transformations and develop a common understanding of the process. These frameworks for discussion or 'frame reflections' (Rein and Schon 1994) may be no more than broadly representative 'working groups' which can be set up to undertake action research and make reports to senior management but have no other authority. They may, however, evolve into reform planning units with considerable institutional authority and dedicated to a process of 'creating institutions that make discussion of what to do inextricable from discussion of what is being done and the discussion of standards for apportioning gains and losses inextricable from apportionment' (Sabel 1994: 231). The project monitoring system may be less a means – either 'quantitative' or 'qualitative' – to manage the ultimate end of 'impacts' or 'transactions'. Rather, the monitoring system must play a major part in establishing the framework of discussions which negotiate the common meanings and resolutions which in turn allow the reinterpretation of positions and the derivation of reassurances about the negotiation and processing of benefit and loss. The hope is that the discursive institutions of the process can evolve piecemeal norms, benchmarks and standards which will allow the positions to be reinterpreted, and reassurances to be derived, about the negotiation of benefit and loss.

The difference in the Russian and Indian cases is that the Russian experiment created and validated a *process monitor* – initially, an individual – as a key factor in the experiment. The Indian forestry experiences have, on the whole, been less fertile grounds for process monitoring either because the monitoring could so easily be interpreted as 'interference' or as 'unnecessary because the executive agency must be left to develop ownership'. The hiatus between the discourse of 'means–ends' and 'solution-side' thinking and the donors' insufficiently anchored intentions to institute flexible planning and 'ownership' was also a cause of the difficulty. At least one of the Indian forestry cases has assumed that 'interpretation/facilitation' and 'institutional development' are essentially parallel, equally technical, skills which could be deployed or taught alongside forest management, agronomy, farm economics etc. But process monitoring is both more and less – it is (vide Vickers in the opening epigram) the *resolution* of specific difficulties and their transformation into norms and the *validation* of those resolutions and transformations into wider arenas. It *is* the project, not part of it in contention for voice and view. An initial decision to set up a validated and independent process monitor allows a variety of resolutions as innovations and uncertainties appear. In the Indian forestry cases the understanding of the need for process monitoring has been uneven. Moreover, it became a special tool of social science and of 'the social faction' rather than an aspect of negotiated multi-disciplinary thinking. In at least the one case reported, the process of developing a discursive institutional framework was threatened and formulae and ends-means menus and deals were increasingly substituted. The overt result has often been the appearance of process factionalism and apparent conflicts between 'people' and 'trees' objectives. In fact, neither the detail nor the substance of people or trees was much in evidence in the projects' discussions, even though they were the ciphers and icons of the struggles. The key tacit issues were usually questions of institutional territory and power and whether or not these could be resolved, modified or appeased through the use of new norms and standards when, that is, the evidence of action research was sufficiently compelling.

The evidence from the two marker country cases is that 'process planning and monitoring' is an important aspect of policy reform but that it should not be left to chance or intention. On the one hand, it must be acknowledged that flexible planning is hard to plan for! Intentions and previous styles of organisational work are what usually sustain the process for individuals and groups. On the other hand, the two country cases suggest that a more explicit theory of the role of process monitoring in policy reform is necessary. We have found the concept of discursive institutions charged with the successive resolution of issues, norms and standards especially useful.

Resolution is, indeed, a very special and useful word. The aim of a process project is to tame controversy through *resolving* the conflicts and clashes of meaning and interest, that is by moving issues from discord to concord by first dissolving and then resolving difficulties. What are needed are institutional *resolvents* – devices selected, that is, to refocus clashes and tensions. Sometime these resolvents are the devices of democracy – extensive discussion, explanation and the mechanism of the vote and debate. At other times the resolvent is the device of the workshop, the planning game, the specially commissioned training or study tour, the special task, or the mediation of the project facilitators to stimulate the refocussing which is necessary. If conditions allow and the environment is tolerant of the creation of new meaning, the resolvent can be the well-chosen metaphor, or the airing and discussion of an encapsulated vision, and so on. The concern should not be, at least in the initial stages, with ends and objectives but with means. Any prior choice of ends will mean the imposition of ends on some categories of stakeholder and the likelihood of overt or covert dissent and opposition. Emphasising the means of resolution through information, through formal decisions (*that* kind of *resolution*) or through statements about general opinion (another kind of *resolution* mechanism) channels dissenting opinion and behaviour towards norms of resolution and helps establish the standards. *Process monitoring* has a key role to play because it is capable of being *resolutory* – that is, explanatory or enlightening. There are no 'quick fixes' or magic solutions. The adoption of an appropriate rate and direction of change and a genuine openness to the diversity of opinion and social standards and aims will allow the early validation of the process monitoring role. Information is power and influence as the Russian example shows – but it is doubtful if much is gained by drawing attention to this since everyone knows it through experience and only questions it when the use of power lacks social authority. Mechanisms of resolution are essential to allow truths to be monitored and documented and for these to be presented to power (Wildavsky 1987). This commitment to speaking truth to power but doing so gradually and through an open-ended *Learning Process* project has a number of important project planning and policy reform implications.

The implications for development projects and policy

The practical implications of our perspective and analyses include considerations for:

- policy dialogue and conditionality;
- process and blueprint planning methods;
- the extent of the policy change ambitions;
- the selection, recruitment and learning of process monitors;
- the development of local action research capacity;
- the nature of the investment in process change.

These implications are taken in turn in the following list of recommendations to process planners and monitors.

The commitment to developing discursive institutions requires the early definition of mechanisms which give a voice to the disenfranchised and vulnerable – either through representatives accountable to those for whom they speak and/or through effective social science research and monitoring. The clients and donors need first to agree to establish these mechanisms. If there is any area in which to lever aid conditionality it is here – in a guarantee that, either directly or indirectly through bottom-up planning and action research, the poorest and most vulnerable stakeholders can make their views known. If negotiated as a precondition of project implementation, the requirement for open discursive institutions and process monitoring may not, as David Mosse fears (this volume), be subsequently dismissed by the executive agency as complicating. The specific results of action research and the forms of consultation and participation may subsequently be contested at the time they are discussed within the project, but the standard of open discussion and a free search for institutional change through agreement and on the basis of evidence must be accepted from the outset.

There are also implications for the contrast between blueprint and process project planning and more generally for project planning methods. The usual criticism of blueprint projects is that they involve limited, non-transformative ends and over-delineated and predetermined means. The process projects reviewed in this chapter are liable to the criticism that for limited ends they substituted generalised and deliberately understated ones and in place of the usual over-specification of means they substituted under-examined, uncertain ones. Logical Framework project planning is fundamental to ODA/DFID operations but may not be as important to other donors. Used too early in ODA projects it can create considerable tension. First, the LogFrame planning, although strategic, takes place in workshops which are indistinguishable from workshops held to discuss far more mundane matters and the result may be to change workshops in general from trusted discursive institutions into battlegrounds to settle the nature of the single project purpose, etc. Second, early LogFrame planning causes tension because it is often unfamiliar to many of the participants, who consequently come to distrust it, its facilitators and its outcomes. Third, it causes tension even between those who are most familiar with the method and its conventions as they experience (*vide* Apthorpe) snappy solution-side diagnosis in the LogFrame sessions rather than hard, field work-based diagnosis of pathology, remedy, prognosis and incremental resolution. Finally since, as we saw in the Russian case, 'impact' hypotheses will be hard to define and test, there will necessarily be a large number of rather key entries in the fourth, 'assumptions' column of the Logical Framework matrix!

The sociologically trained observer can usually find interest and further points for influence in understanding the discursive hiatus between ends–means planning and the learning process. But it can be disconcerting for the project staff and administrators who are charged with managing the projects and sanctioning the monitoring arrangements. The resolution of these gaps in knowledge and the potential elaboration of new common understandings and new common conflicts is a key part of process monitoring and the discursive institutions that need to be established. The uncertainties in information and purpose are, however, considerable and can lead to severe stress on the part of managers. This stress and tension is severely threatening to the change process itself which requires toleration of many stakeholders' views and difficulties. It is recommended, therefore, that senior managements do not set impossibly high goals and that they accept the long time horizons and falterings in the process of transformation.

The process planner and monitor needs independence, continuity and the opportunity to learn. It is a distinctive kind of independence. First, it must be capable of moving from (*vide* Apthorpe again) 'hard' academic analysis of discourse and transaction to the therapies of resolution and validation; neither imprisoned, that is, in the officer/official's public donative discourse nor trapped in the social scientist's marvelling at the arabesques of institutional behaviour and pathos. Second, it is best found independently of technical disciplines – it almost requires someone who is homeless in terms of academic disciplines. Third, once found the trusted process planner/monitor for Project 'X' should not be changed to suit competitive bidding requirements, changing staff resources or other external pressures. S/he is researching, learning about and recording a highly exposed and extended institutional ethnography, building rapport and understanding and needs the time to dig, build trust and understand the complexities and room for manouevre on behalf of change. S/he is also learning new techniques and methods to deal with the challenges of the project or policy area and needs space and time to do so.

At the same time as the process planner/monitor is given time and space, the project needs to develop local action research capacity to preserve the independence and usefulness of the monitoring after the initial stages. This has become an important requirement for the future of the Russian case and is being addressed, with varying success and approaches, in the Indian forestry cases.

Vinay Luthra (1995) in the papers for the contributing conference raises questions about investment in skills or systems in the case of the Western Ghats forestry project in Karnataka. He notes that it was individuals and social networks that were working well in his example, not systems. The implication is that process projects must first find ways of investing in human resources and specific people and move to systems later. This means tackling the problem of staff transfers and assessing the risks and chances of projects that are 'transfer-tolerant' (Farrington, Gilbert and Khandelwal, this volume). Early attention is needed in process planning projects for training in adaptive or contingency planning, for problem-solving skills, and for generalised interpretative and process therapy skills. If enough staff from the executive agency and partner institutions were trained in these generic skills early on in the process, the state or province could continue to access the skills in locations and subjects outside the immediate pilot project and beyond the immediate system and procedural changes introduced.

In summary, we note that in policy process or project changes the *resolutive* must take pride of place. It is usually the individuals and the discursive institutions centred on social networks that innovate and stay in the frame while systems and measures prove difficult to define exactly or to implement. If fortune smiles on the process changes there may come the chance for the next stage of formal *resolutions*, either in the sense of formal decisions and regulations of a public body – this is the way that Poffenberger (1990) uses the word – or as majority and somewhat binding expressions of opinion in a formal meeting of people and their representatives. Even though this stage will seem like a 'real' policy change compared to the earlier discursive stage, both are in fact necessary and the process planner and monitor needs to pass backwards and forwards between the two. Finally, or in parallel, there may be the *resolute*, whether a resolute individual or a determinative purpose, the vision. If this person or vision is available and can be maintained then the process will accelerate. If, however, project activities do not lead to resoluble outcomes then the resolute are unlikely, in our pluralistic environment, to solve the difficulties although they will fight their battles no doubt. It may be, however, that the investment funds are best spent elsewhere, in places where resolution is possible.

Note

1   The Know How Fund is Britain's programme of bilateral assistance to the countries of central and eastern Europe and central Asia. Its aim is to support the process of transition in a way which promotes and recognises the rights and

interests of all people. It is administered by the Department for International Development.

References

Apthorpe, R. (1986) 'Development policy discourse', *Development and Change* 6: 377–89.

Beteille, A. (1996) 'Epilogue: village studies in retrospect', in *Caste, Class and Power: Changing Patterns of Stratification in a Tanjore Village*, 2nd edn, Delhi: Oxford University Press, pp. 231–51.

Blair, H. (1996) 'Democracy, equity and common property resource management in the Indian sub-continent', *Development and Change* 27, 3: 475–527.

Breman, J. (ed.) (1997) *The Village in Asia*, Delhi: Oxford University Press.

Brustinow, A. and Turner, R. (1996) *Russia Farm Privatisation*, Swansea, mimeo.

Chambers, R. (1994) 'The origins and practice of participatory rural appraisal', *World Development* 22, 7: 953–69.

Crozier, M. (1964) *The Bureaucratic Phenomenon*, London: Tavistock Press.

Escobar, A. (1995) *Encountering Development: the Making and Unmaking of the Third World*, Princeton, NJ: Princeton University Press.

Eyben, R. and Ladbury, S. (1995) 'Popular participation in aid-assisted projects: why more in theory than in practice', in N. Wright and S. Nelson (eds), *Power and Participatory Development: Theory And Practice*, London: Intermediate Technology Publications, pp. 192–200.

Gregory, R. (1989) 'Political rationality or 'incrementalism'?: Charles E. Lindblom's enduring contribution to public policy making theory', *Policy and Politics* 17: 139–54.

Hill, M. (ed.) (1993) *The Policy Process: a Reader*, London: Harvester Wheatsheaf.

Hobley, M. (1996a) *Participatory Forestry: the Process of Change in India and Nepal: Rural Development Forestry Study Guide 3*, London: Overseas Development Institute.

—— (1996b) 'Institutional change within the forestry sector: centralised decentralisation', *ODI Working Paper No. 92*: 3–49.

Kant, S., Singh, N.M. and Singh, S.K. (1991) *Community Based Forest Management Systems (case studies from Orissa)*, Bhubaneswar: ISO/SwedeForest, IIFM, SIDA.

Lindblom, C.E. (1959) 'The science of "muddling through"', *Public Administration* 19: 79–99.

—— (1980) *The Policy-Making Process*, Englewood Cliffs, NJ: Prentice Hall.

Locke, C. (1995) 'Planning for the participation of vulnerable groups in communal management of forest resources: the case of the Western Ghats forestry project', PhD Thesis, University of Wales, Swansea.

Luthra, V. and Rees, J. (1995) 'The Western Ghats – monitoring a process project', paper prepared for ODI/CDS workshop on The Potential for Process Monitoring in Project Management and Organisational Change: Lessons from the Natural Resources Sector, April 1995.

Overseas Development Administration (ODA) (1996) *Sharing Forest Management: Findings from ODA's Review of Participatory Forest Management*, London: Overseas Development Administration.

Palit, S. (1996) 'Indian Forest Departments in Transition', in M. Poffenberger and B. McGean (eds), *Village Voices, Forest Choices: Joint Forest Management in India*, Delhi: Oxford University Press, pp. 210–29.

Peattie, L.R. (1994) 'Society as output: exit and voice among the passions and interests', in L. Rodwin and D.A. Schon (eds) *Rethinking the Development Experience: Essays Provoked by the Work of Albert O. Hirschman*, Washington DC: Brookings Institute.

Poffenberger, M. (ed.) (1990a) *Keepers of the Forest: Land Management Alternatives in Southeast Asia*, West Hartford, CN: Kumarian Press.

—— (1990b) 'Facilitating Change in Forestry Bureaucracies', in M. Poffenberger (ed.), *Keepers of the Forest: Land Management Alternatives in Southeast Asia*, West Hartford, CN: Kumarian Press, pp. 101–18.

Poffenberger, M. and McGean, B. (eds) (1996) *Village Voices, Forest Choices: Joint Forest Management in India*, Delhi: Oxford University Press.

Poffenberger, M., McGean, B. and Khare, A. (1996) 'Communities sustaining India's forests in the twenty-first century', in M. Poffenberger and B. McGean (eds), *Village Voices, Forest Choices: Joint Forest Management in India*, Delhi: Oxford University Press, pp. 17–48.

Poffenberger, M. and Singh, C. (1996) 'Communities and the state: re-establishing the balance in Indian forest policy', in M. Poffenberger and B. McGean (eds), *Village Voices, Forest Choices: Joint Forest Management in India*, Delhi: Oxford University Press, pp. 56–85.

Rein, M. and Schon, D.A. (1994) *Frame Reflection: toward the resolution of intractable policy controversies*, New York: Basic Books.

Rew, A. (1997) 'The Donors' Discourse: official social development knowledge in the 1980s', in R. Grillo and R.L. Stirrat (eds), *Discourses of Development: Anthropological Perspectives*, Oxford: Berg Publishers.

Rew, A., Saxena, N.C., Guhathakurta, P., Misra, P.K., Shrivastava, R. and Sjoblom, D. (1995) 'A Pre-Appraisal of the Orissa Forestry sector Development Programme', Delhi: SIDA.

Rodwin, L. and Schon, D.A. (eds) (1994) *Rethinking the Development Experience: Essays Provoked by the Work of Albert O. Hirschman*, Washington DC: Brookings Institute.

Rondinelli, D.A. (1983) *Development Projects as Policy Experiments*, London: Methuen.

Sabel, C.F. (1994) 'Learning by Monitoring: the institutions of economic development', in L. Rodwin and D.A. Schon (eds) *Rethinking the Development Experience: Essays Provoked by the Work of Albert O. Hirschman*, Washington DC: Brookings Institute.

Smith, G. and May, D. (1980) 'The artificial debate between rationalist and incrementalist models of decision-making', *Policy and Politics* 8: 147–161.

Sood, M.P. (1996) 'New forestry initiatives in Himachal Pradesh', *IIED Forest Participation Series* 3: 1–20.

Uphoff, N. (1992) *Learning from Gal Oya: Possibilities for Participatory Development and Post-Newtonian Social Science*, Ithaca, NY: Cornell University Press.

Vanaik, A. (1990) *The Painful Transition: Bourgeois Democracy in India*, London: Verso.

Wildavsky, A. (1987) *Speaking Truth to Power: the Art and Craft of Policy Analysis*, New York: Transaction Books.

Wright, S. and Nelson, N. (eds) (1994) *Power and Participatory Development: Theory and Practice*, London: Intermediate Technology Publications.

BIBLIOGRAPHY

Workshop papers: the potential for process monitoring in project management and organisational change: lessons from the natural resources sector, 6–7 April 1995

Farrington, John. 'ODI. A background note on process monitoring and documentation'.

Rew, Alan. 'Development plans, processes and resolvents: issues in the analysis and reporting of institutional change in rural areas'.

Govan, H. 'Incorporation of a participatory approach into a coastal aquaculture research programme'.

Kleitz, Gilles. 'Learning needs and fields of possible contribution'.

Lewis, David 'From "partnership on paper" to "partnership in practice": achieving course corrections on ICLARM's aquaculture project in Bangladesh'.

Mosse, David. 'Process documentation and participatory rural development'.

Norrish, Pat. 'Experience in South Africa: process monitoring of process'.

Quan, Julian. 'Process monitoring and documentation: a note on NRI Social Development Section experiences and learning requirements'.

Salomon, Monique L. 'RAAKS: a process approach to innovation'.

Sotomayor, Octavio and Bebbington, Anthony. 'The relationship between NGOs and governments in agricultural research and extension: taking a look at policy processes'.

Brustinow, Angelika. 'Notes on process project monitoring'.

Clarke, Gerard. 'Process Monitoring: mitigating the inter-institutional pressures in development projects.

Luthra, Vinay and Rees, Julia. 'The Western Ghats: monitoring a process project'.

Montgomery, Richard. 'General issues; specific comments'.

Davies, Rick. 'An evolutionary approach to facilitating organisational learning; an experiment by the Christian Commission for Development in Bangladesh (CCDB) with the People's Participatory Rural Development programme (PPRDP)'.

195

INDEX

Milton Keynes UK
Ingram Content Group UK Ltd.
UKHW031531071024
449327UK00005B/128